REDISCOVERED CLASSICS

AHEAD OF ITS TIME

THE EARLY TWENTIETH CENTURY ELECTRIC VEHICLE

A STANDARD HANDBOOK FOR ELECTRICAL ENGINEERS

BY JOHN G. CAROLL & WALTER E. WINSHIP, PhD

WITH A NEW FOREWORD BY NICO HOERNLE

ENDLESS MOUNTAI
PUBLISHING COMPANY
MOOSIC, PENNSYLVA

AF869994

Copyright © 2019 Endless Mountains Publishing Company

All rights reserved.
No part of this publication may be
reproduced or transmitted in any form or by any means,
without permission in writing from the publisher

ISBN: 978-0-9988527-4-4

This is a reproduction reprinted from
The Standard Handbook for Electrical Engineers
Written by John J. Carroll and Walter E. Winship, PhD
Edited by Frank F. Fowle, S.B.
Originally published by McGraw-Hill Book Company, 1922

With a new foreword by Nico Hoernle

Published by Endless Mountains Publishing Company
Moosic, Pennsylvania
www.EndlessMountainsPublishing.com

MANUFACTURED IN THE UNITED STATES OF AMERICA

FOREWORD

Few inventions have as colorful a history and held such promise to disrupt a major commercial sector as electric cars. The electric car has captured our imagination for nearly two centuries, outlasting many other products and technical ideas. It's an invention characterized by enormous uncertainty, idealistic goals, fierce competition that almost led to its demise, and an improbable resurgence. As they have in the past, electric vehicles (EVs) have once again begun to compete head on with the gasoline- and diesel-fueled internal combustion engine car. While EVs never enjoyed the same success as their internal combustion engine rivals, environmental concerns, such as noise, pollution and global warming, combined with leaps in battery technology could tilt the balance in their favor. It is therefore interesting to look at the evolution of EVs and to study their history to see what engineering and design shortfalls they struggled with in the past and how modern solutions have propelled EVs into the future.

Ahead of its Time: The Early Twentieth Century Electric Vehicle was a handbook written for and by electrical engineers in 1922, describing commercially available EVs at the time. When it was written, EVs maintained a respectable 30% market share, holding their own against gasoline and steam-powered vehicles. In fact, at the end of the 20th century, it wasn't clear which technology would win out as the backbone of the transportation sector. Unlike their gasoline-fueled counterparts, EVs start instantly, are quiet, produce no noxious fumes, and have much simpler transmissions. They also possess fewer reciprocating parts, lower fuel costs, faster acceleration, and maximum torque available instantaneously. While steam engines were used in factories and railroads, their weight, relentless thirst for water, and long warm-up period made them impractical for use in cars.

EVs found a home in cities where they served as taxis, vans, trucks, buses, and even private cars in the rare instances when individuals owned a car. Their largest drawback, however, was their limited range, which typically topped out around 60 to 90 miles per charge of their lead-acid batteries. As Americans increasingly hit the road in privately owned cars, seeking to explore their country through newly constructed highways, this lack of range became a larger disadvantage. The advent of cheap and seemingly ubiquitous oil further undermined sales of electric cars. When Ford rolled out the mass-produced Model T in 1908, automobiles became accessible to middle-class Americans, and the internal combustion engine vehicle took off as the affordable and popular option. And when, the electric-starter for internal combustion engines was invented in 1911—rendering the dangerous, difficult and inconvenient process of hand-cracking a thing of the past—that innovation dealt another large blow to EVs. Drivers no longer had to worry their engine might kick back, causing the crank to jerk and reverse direction, possibly injuring the operator, e.g. breaking his or her arm. Suddenly, internal combustion cars were almost as easy to start as EVs, and after 1935, EVs all but vanished from our roadways. Occasional crises such as the 1973 oil embargo put EVs back on the table, but attention waned again as the turmoil faded.

Today, we face a larger crisis, and the urgent need for vehicles with minimal environmental impact has necessitated innovation in the transportation sector. As a result, with their promise for drastically lower emissions (particularly of CO2), EVs have undergone a transformation and seen a dramatic resurgence in popularity.

Surprisingly, the original concept of an EV remains relatively unchanged and modern electric cars rely on many of the same components that EVs used over a hundred years ago, although now with a modern twist. For instance, lithium-ion batteries have neatly addressed one of their main historic shortcomings, adding hundreds of miles per charge. While today's EVs still struggle with charge times of about 20 minutes to hit 50% and more than an hour to reach 100% (even in the best-case scenario), improvements in this arena are the next hurdle to jump.

With a little creativity, it's not hard to imagine how much further EV technology can go in today's society. The EV's electric system naturally lends itself to incorporating autopilot. At first, this functionality will likely allow the vehicle to drive autonomously on highways. Later, the autopilot will be able to park the car autonomously and the driver will summon the vehicle from the parking place. Before long, EVs could become fully autonomous, ferrying passengers around with the touch a button. Such advancements would have huge implications for vehicle ownership and transportation. Autonomous EVs might travel in dedicated lanes at greater speeds, not limited by human reaction time but rather by the speed and processing power of computers. In addition, EVs could communicate with each other wirelessly and one day make many traffic lights obsolete by timing their arrival at intersections perfectly with other traffic.

Far from a modern marvel, EVs were some of the first cars on the road. Studying their history allows us to understand their evolution, gives insight into their present-day design, and allows us to learn from past mistakes. Moreover, it is fascinating to imagine where they might take us next.

—Nico Hoernle, 2019

ELECTRIC VEHICLES

by John G. Carroll, 1922

Chief Engineer, Walker Vehicle Company
Member of the America Institute of Electrical Engineers and
Society of Automotive Engineers

ELECTRIC VEHICLES

GENERAL CLASSIFICATION

1. The electric vehicle art prior to 1910 exhibited relatively little progress, due partly to the poor design of the earlier types of vehicles and partly to the comparatively heavy unsuited batteries. The art began to show considerable activity in 1910 and for the following four years the development of the passenger car progressed more rapidly than that of the commercial truck. However, the progress in the design and development of the gasoline passenger car, with its electrical starting and lighting system, higher speeds and practically unlimited radius of action, resulted in the diversion of popular attention from the electric passenger car to the gasoline passenger car. For this reason the number of manufacturers decreased steadily, until in the beginning of 1921 there were only two prominent makers of electric passenger cars.

2. Progress in commercial and industrial vehicles. The electric commercial and industrial trucks and tractors have progressed steadily since 1910 and have proved their economic worth in their particular fields in the transportation of merchandise.

3. Classification. Electric vehicles may be subdivided into three general classes according to character of service:

(1) Passenger cars.
(2) Commercial trucks and tractors.
(3) Industrial trucks and tractors.

The source of energy supply permits another classification, namely storage-battery vehicles and the so-called trackless trolley which is a motor-propelled vehicle receiving its energy from an overhead trolley system. The storage-battery vehicle is by far the more important at the present time, as the highway trolley vehicle has not yet come into extended use in this country although it is employed to a limited extent abroad.

4. Scope of presentation. Due to the broad scope of design, the subject of electric vehicle engineering can be presented in the brief space available in an engineering handbook only with difficulty. An attempt has been made in this section to embody merely the fundamentals of the subject based on modern practice and so far as possible describe the most representative vehicle practices obtaining at present. Because of the variety of designs now being produced, it is impracticable to describe them all. Numerous types of vehicles have been designed for special purposes such as passenger and baggage transfer, cable pulling, crane tractors, pole hauling, dumping, baggage trucks, fire apparatus, etc., but practically all of these applications are specific and no attempt has been made to include them.

5. Features emphasized. It is particularly desired to point out that where descriptions of machines and accessories of various makes have been omitted, such omissions are due to the fact that limitations of space have confined the subject matter to those types which have gained prominence in industry or whose designs embody characteristics of particular interest to the engineer.

PASSENGER CARS

6. General features. The passenger vehicle involves no engineering features which do not conform to the fundamentals of vehicle engineering and their applications to commercial trucks as discussed under that caption.

7. Chassis and body. All cars are comprised of two principal elements; the chassis and the body. In general, the chassis comprises the frame, front and rear axles, complete with wheels, springs, steering gear, motor, power-transmission members, controller and battery. The plan view of a typical electric passenger-car chassis is shown in Fig. 1.

8. The chassis frame is usually of pressed steel channel sections of liberal depth, with cross members for supporting the motor, battery compartment and other equipment forming a part of the vehicle.

9. The front axle is of drop-forged I-beam construction, providing maximum carrying capacity with minimum weight. The **steering knuckles** are, in general, drop-forged from alloy steel and heat treated.

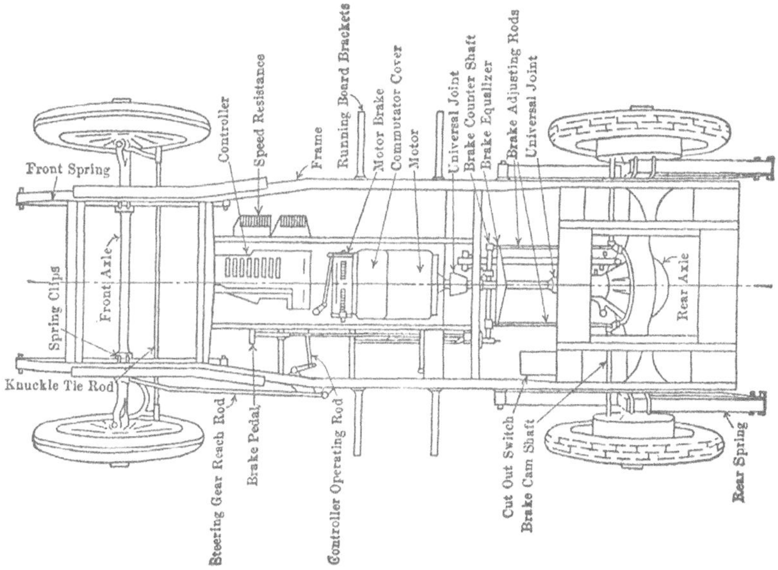

FIG. 1.—Plan view of Detroit electric chassis.

10. Power-transmission systems have in the past embodied various combinations of chain, spur and bevel gear, and worm drive. In the present-day designs, the rear axle is generally of the three-quarter floating type having either spiral bevel gears, with an approximate reduction of 5 to 1, or worm gearing with an approximate reduction of 9¾ to 1. Power is usually delivered from the motor to the rear axle through a propeller shaft having two dust-proof universal joints. The rear-axle torque is usually carried through a pressed steel member, pivotedly supported, to the chassis frame.

11. Two separate sets of brakes are usually provided, operating independently. The two sets are usually of one of the following combinations: (*a*) External contracting and internal expanding type mounted on the rear wheels and operated by foot pedals; (*b*) internal expanding type mounted on the rear wheels, operated by a foot pedal, and external contracting type mounted on the front end of the motor, fitted onto the armature or propeller shaft and operated by a backward movement of the controller lever or by a foot pedal. In some designs the application of the service brakes simultaneously interrupts the power circuit regardless of the position of the controller lever, making it impossible to start the car again until the controller lever has been returned to its neutral position.

Dynamic and regenerative braking have been tried in the past, but neither is in general use at present due to the complications that are introduced into the control system.

12. The springs are generally of chrome-vanadium heat-treated steel, the front being of the semi-elliptic type and the rear either the three-quarter elliptic or cantilever type.

13. The wheel base on cars of present-day design is usually 100 to 105 inches.

14. The wheels are usually quick-detachable wood or wire type, designed for approximately 32 × 4½ in. cord pneumatic tires.

15. Batteries. Two prominent manufacturers specify their standard battery equipment to provide for 40 to 42 cells of 13 to 15 plate lead-acid battery. Vehicle batteries are discussed more fully in Par. **128**.

16. The motor is usually a 4-pole series-wound machine, liberally designed to insure efficiency and ability to withstand overloads encountered under all operating conditions. The armature bearings of the motor are of the ball type. The motor is usually supported from cross members of the chassis frame. See Par. **87**.

17. The controller is usually of the drum, continuous-torque type providing approximately five or six speeds forward and three or more reverse. The controller is generally supported from the chassis frame and actuated by a lever mounted beside the steering lever. The controller is discussed more fully in Par. **108**.

18. The operating levers for steering and control are usually mounted at the side of the seat, in parallel position one above the other. The steering lever, usually the longer, is operated by the right hand and the controller lever, usually the shorter, by the left hand.

19. The radius of operation is usually from 60 to 90 miles per charge of battery, depending on topography, road conditions and the speed at which the car is operated.

COMMERCIAL TRUCKS AND TRACTORS

20. Classification. Commercial trucks may be divided broadly into four classes:

(1) Those having electric propulsion, the energy being furnished by a battery.

(2) Those propelled by internal combustion engines.

(3) Those propelled by steam.

(4) Those having gas-electric or steam-electric propulsion in which the energy of the prime mover is electrically transmitted to the axle or wheel.

21. Economic field of application. Each type of commercial truck has its distinct field of economic application, that of the electric vehicle being found in city delivery embracing the interchange of merchandise between railway terminals and docks, stores and factories, and retail deliveries. The electric vehicle possesses superior advantages over other means of merchandise transportation due to its inherent characteristics of rapid acceleration and retardation permitting it to maintain a high average speed and due further to its freedom from intricate mechanism for transmitting the motor energy to the wheels and the absence of numerous reciprocating parts.

22. Economic field of gasoline vehicle. The gasoline motor vehicle finds its best field of application in uninterrupted long-haul deliveries where the mileage exceeds that safely and conveniently obtained from the electric truck, its economic value being governed entirely by the individual characteristics of the particular business to which it is applied.

23. Economic field of electric vehicle. Electric trucks are more economical and satisfactory than gasoline trucks for approximately 85 per cent. of all city haulage. The salient characteristics which contribute to the high efficiency and low operating costs of electric trucks are:

(*a*) The low cost of electrical energy.

(*b*) Operation at a safe economical speed.

(*c*) Ease of maintenance of battery service.

(*d*) Reliability and uniform driving power of the electric motor, which delivers maximum power to the rear wheels.

(*e*) Substantial construction and elimination of reciprocating members, which almost wholly eliminate road troubles, reduce maintenance expense to a minimum and obviate the necessity for high-salaried mechanics.

(*f*) Minimum vibration and prevention of careless speeding prolong the life of the electric truck, and make it more than double that of a gasoline truck.

(*g*) The motor draws no energy from the battery when the brakes are applied or when the truck is coasting.

(*h*) Low cost of insurance.

24. The chassis of the various makes differ only in some special features of design. The main frame members are comprised of rolled or pressed steel channels, with cross members secured to the side sills with either proper angle iron and gusset plates or integral gusset plates.

25. Representative types of chassis are illustrated in Fig. 2 and 3, showing respectively the side elevation and plan of a Ward one-ton chassis and a Walker 3½-ton chassis. Both of these types employ a single motor, the former with a worm-drive axle and the latter with a special balanced drive as described more fully in Par. **32.**

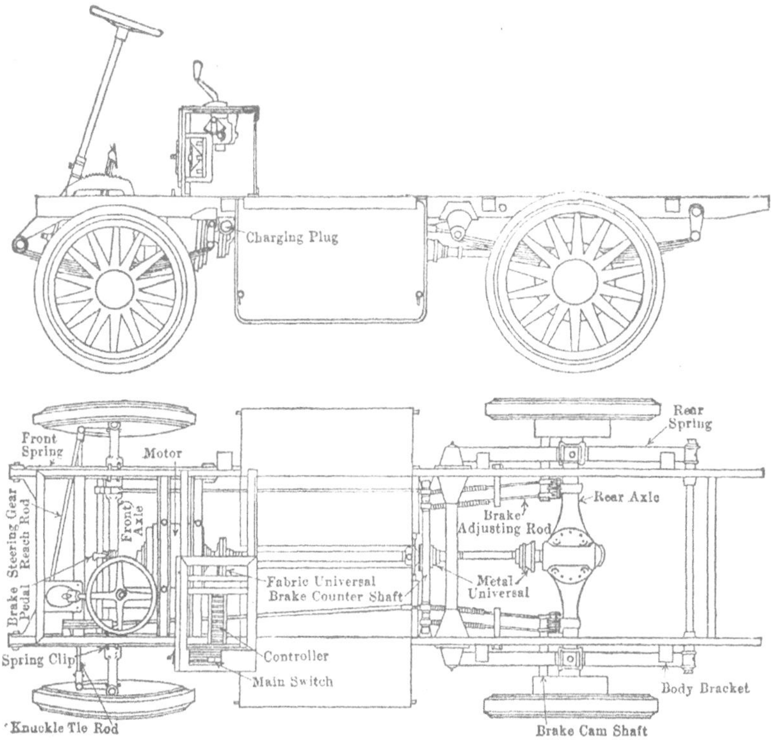

FIG. 2.—Elevation and plan view of 1-ton Ward electric chassis.

26. The battery compartment is usually carried amidship by structural steel members hung from the main chassis frame and having truss rods from the lower corners of the compartment to the side sills of the frame. In some cases the battery is mounted on top of the side frames and partly under the driver's seat. The underslung type of battery compartment has the following advantages: (*a*) Low center of gravity, resulting in less tendency when turning corners for the weight to be thrown from the inside to the outside wheels and therefore less tendency to skid; (*b*) the overall length of the truck is reduced, which enables the truck to be more easily operated in congested traffic zones and narrow places and requires less garage space; (*c*) permits the use of removable battery cradles and facilitates the interchange of batteries.

27. Springs. The front and rear springs are usually of the semi-elliptic type, of high-grade alloy steel, heat treated.

28. Axles. The front axles are usually of I-beam section, drop forged from high-grade steel, with drop-forged steering knuckles and steering knuckle arms. The rear axles embodying the transmission of the motor energy to the wheels differ in design on all the different makes of trucks.

29. Transmission systems embody various combinations of chain, spur gear, bevel gear and worm drives. Considerable difference of opinion exists as to what constitutes the most desirable method of motor application and the best transmission design. The essential elements of a transmission comprise means of reducing the high speed of the motor to the comparatively

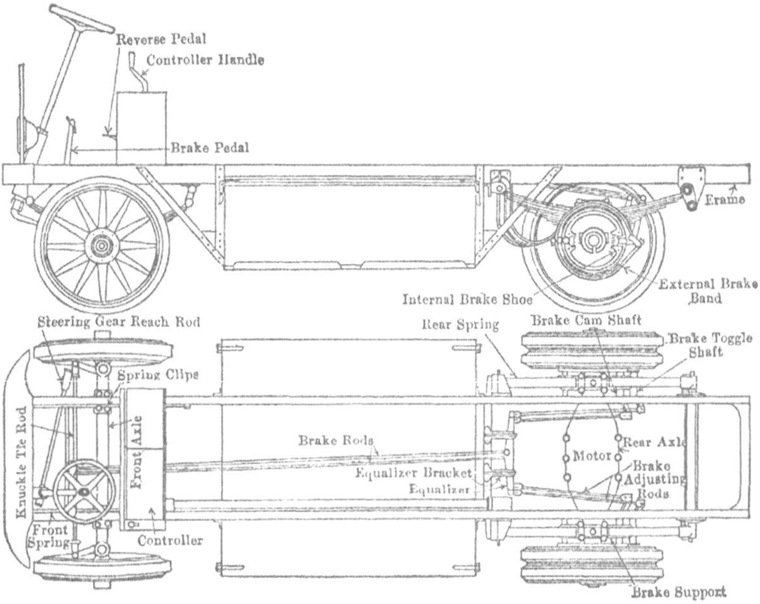

FIG. 3.—Elevation and plan view of $3\frac{1}{2}$-ton Walker electric chassis.

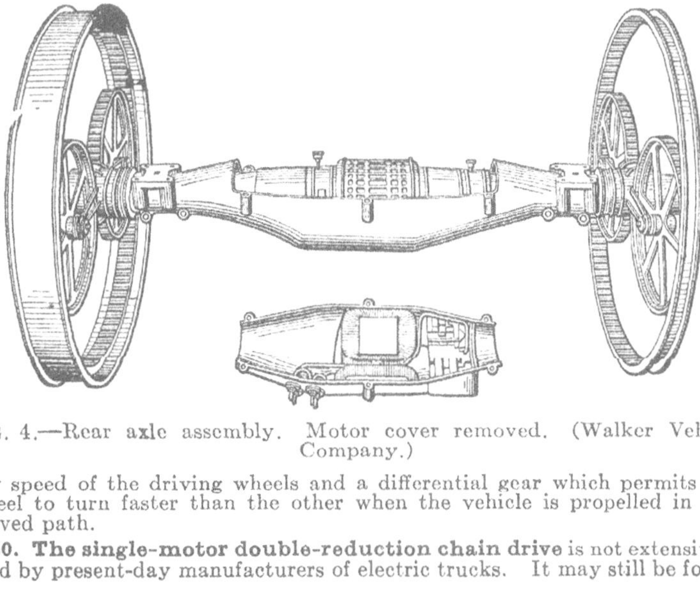

FIG. 4.—Rear axle assembly. Motor cover removed. (Walker Vehicle Company.)

low speed of the driving wheels and a differential gear which permits one wheel to turn faster than the other when the vehicle is propelled in any curved path.

30. The single-motor double-reduction chain drive is not extensively used by present-day manufacturers of electric trucks. It may still be found in operation on old vehicles in service in this country and on several makes of electric trucks manufactured in this country and in England.

31. The worm drive is usually of the semi-floating or full-floating type having the worm, worm wheel and differential and their bearings assembled as a unit with the cover. The housing is usually of cast steel reinforced by tubes. The road wheel is driven through a squared shaft and driving flange bolted to the hub, or through a splined shaft in the hubs of the differential gear and driving flange bolted to the wheel hub.

32. Balanced drive. In the Walker balanced drive, the motor, differential and transmission members are all embodied as parts of the rear axle and wheel assembly (Fig. 4). The power is transmitted through drive shafts to floating pinions within each wheel, then through intermediate spur gears housed within the disc wheels, to the internal gear which is pressed into and keyed to the wheel bands or rims. The drive pinions and shafts have no bearing, the pinions floating in a balanced manner between the idler gears with which they are constantly in mesh. The rear wheels are dirt- and water-proof and are partly filled with lubricant which is distributed continuously over all working parts. The rear wheels are mounted on roller or ball bearings supported by drop-forged steel stub axles secured in the ends of the hollow rear axle. The mechanical efficiency of this drive is very high, and stand tests indicate that efficiencies as high as 98 per cent. are obtainable.

33. The wheel-drive systems embodying constructions in which a motor drives each wheel separately and forms a more or less self-contained unit therewith are described in Par. **34** and **35.**

34. The couple gear drive, shown in Fig. 5, employs a motor mounted in a fixed horizontal position within the wheel itself, the axle stubs on which

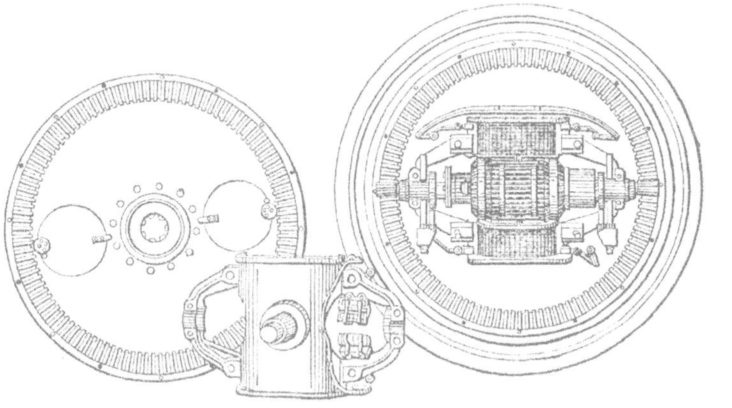

Fig. 5.—Wheel assembly, couple gear.

the wheel turns being integral with the motor castings. The armature shaft makes a slight angle with the plane of the wheel, permitting the bevel pinions mounted on each end of the armature shaft to mesh with their respective halves of a double cog rack near the periphery of the disc wheel. Thrust stresses arising from one bevel set are balanced by the other. This arrangement lends itself to four-wheel drive and steering.

35. Other wheel-drive systems which obviate the use of bevel gears and embody the double spur-gear reduction are shown in Figs. 6 and 7. These drives are used by the Commercial Truck Company. In the concentric gear drive illustrated in Fig. 6, the gear cases are mounted between two axle members, the extension of the case forming the axle spindle. The motor is bolted to the inner side of the gear case in a fixed position and in line with the axle. The armature pinion engages three spur gears, which are carried by studs mounted integral with the driving shaft. The pinions which are integral with the intermediate gears, as shown in Fig. 6, engage a single gear which is pressed into and keyed to the gear case. The driving shaft passes through the axle spindle and connects with the outer end of the wheel hub through the hub cap. The driving shaft has no bearing in the axle spindle, the driving gears and driving shaft being supported and held in proper position in the following manner: The double intermediate gear has a cylindrical extension formed on its inner end having a diameter equal to the pitch diameter of the small gear. A cylindrical track is formed in

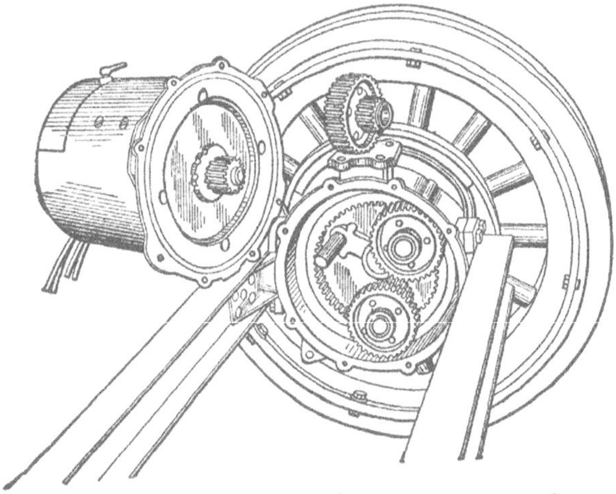

FIG. 6.—Concentric gear drive. (Commercial Truck Company.)

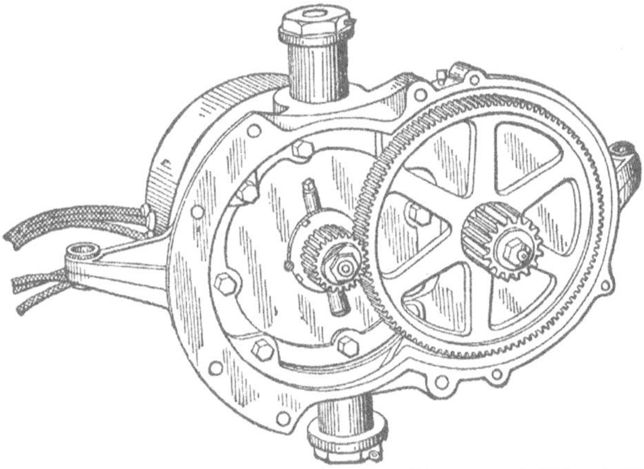

FIG. 7.—Double spur-gear reduction drive. (Commercial Truck Company.)

the motor head, the diameter of which is equal to the pitch diameter of the internal gear. It will thus be seen that the three bearings form a rolling support for the inner end of the driving system, the outer end of the driving shaft being supported and held centrally by the hub shaft. The gear case is oil and dust tight, the gears running continuously in a bath of oil. This drive is used on trucks of ½-ton, 1-ton and 2-ton capacity.

In the double-reduction spur gear drive illustrated in Fig. 7 the motor, which is a standard truck type, series wound, with a cylindrical steel frame, is carried in a steel casting accurately bored to a size that will allow the motor to be slipped freely in place up to a definite shoulder. The case is split partly around and horizontally, so that by tightening bolts through the lugs cast on this part of the case, the case can be securely clamped around the outside of the motor. The gearing between the motor and the wheel consists of straight spur gears. The first pinion is mounted directly on the tapered armature shaft with a key. The double intermediate gear is carried by a roller bearing on an intermediate stud, which is supported in the same casting that carries the motor. The intermediate pinion engages an internal gear mounted directly on the wheel hub, the outside surface of the internal gear forming the brake drum. The axle spindle is constructed so as to form an extension of the gear and motor case, the spindle being approximately in line with the center of the motor. A concentric flange, which is bored to a diameter slightly larger than the outside of the internal gear and provided with grooves, forms a dust and grease type joint. Trunnions integral with the gear and motor case located at the top and bottom, as shown in Fig. 7, are arranged to engage the axle ends and form a knuckle construction for steering the motor wheel as a unit. The gearing is entirely enclosed and the gear case is partly filled with a heavy grease. The axles are steel forgings and are made in two parts separated by a spacer forming the equivalent of a deep I-beam section. The axle forgings are enlarged at the ends and bored out and bushed to receive the trunnions before described, which are part of the motor casing. The axle with the motor casing thus forms a knuckle construction for steering, a thrust ball-bearing being mounted in the front top axle at the trunnions. The assembly of the front and rear constructions is the same, with the exception of the ball bearings. The motor wheels of the front construction are connected by the usual cross rod with the steering rod or drive link connecting one of the motor wheels to the steering gear. The motor wheels of the rear construction are secured in a fixed relation to the axle, the wheels being rigidly mounted. This drive is used on trucks of 3½- and 5-ton capacity.

36. The methods of providing for torsion and propulsion stresses in electric trucks follow the same general practice which prevails in the design of gasoline trucks. The discussion of these methods cannot be undertaken within the space available, and therefore the reader is referred to any standard treatise on the design of gasoline motor trucks.

37. The brakes in general conform in design to the accepted practices of gasoline motor truck brakes. Two independent brake systems are usually furnished, one being an external contracting-type brake on the rear wheel functioning as a service brake and the other being an internal expanding brake on the rear wheel functioning as an emergency brake. In some cases both brakes are of the internal expanding type on the rear wheel and in other cases an external contracting-type brake on the propeller shaft is furnished in addition to an internal expanding-type brake on the rear wheels. The brakes are generally operated by foot pedals.

38. Battery requirements. A storage battery to fulfill the requirements of truck propulsion should have the following characteristics: (*a*) The ability to propel a loaded truck up the gradients encountered in service; (*b*) it should be efficient; (*c*) it should be rugged and substantial in construction and capable of withstanding the severe shocks of service; (*d*) it should have the ability to give a long period of useful service; (*e*) it should have ample capacity for the application involved.

The recommendation of the Society of Automotive Engineers regarding number of battery cells is as follows: The number of cells in lead-acid batteries shall be 42, and the number of cells in the nickel-iron alkaline type 60. In many cases 44-cell batteries cannot be properly charged on account of the low voltage of the current supply. The 42-cell battery can be divided to make a more convenient arrangement of trays. See Par. **128** *et seq.*

39. Motors. One motor of the four-pole series-wound type is usually employed to propel a truck, although in some applications two or four motors are used. Motors are discussed more fully under Par. **87** to **107**.

40. Controllers are discussed more fully under the caption "Controllers," Par. **128** to **132**.

41. Tires. The function of the tires is to provide traction, prevent slippage, protect the mechanism of the vehicle from shock and vibration and cushion the load. In general all electric trucks are equipped with solid tires. The carrying capacity of solid tires as specified in the standards of the Society of Automotive Engineers is given in Par. **42.**

42. Sizes and Carrying Capacity of Solid Tires
(S. A. E. Standard)

Nominal tire size		Min. total sectional area of rubber (1) (sq. in.)	Carrying capacity (lb.)
In.	Mm.		
32 × 3	75/660	5.00	1,000
36 × 3	75/762	5.00	1,000
32 × 3½	90/660	6.75	1,300
34 × 3½	90/711	6.75	1,300
36 × 3½	90/762	6.75	1,300
32 × 4	100/660	7.75	1,700
34 × 4	100/711	7.75	1,700
36 × 4 (2)	100/762	7.75	1,700
34 × 5	125/711	10.75	2,500
36 × 5 (2)	125/762	10.75	2,500
40 × 5 (3)	125/864	10.75	3,000
34 × 6	150/711	13.75	3,500
36 × 6	150/762	13.75	3,500
40 × 6 (3)	150/864	13.75	4,000
34 × 7	175/711	16.75	4,500
36 × 7	175/762	16.75	4,500
40 × 7 (3)	175/864	16.75	5,000
36 × 8 (4)	200/762	19.75	5,500
36 × 10 (4)	250/762	25.75	7,500
40 × 10 (4)	250/864	25.75	8,000
40 × 12 (4)	300/864	31.75	10,000
40 × 14 (4)	350/864	37.75	12,000

(1) Includes both hard and soft rubber.
(2) Single tires for single and dual wheels.
(3) Dual tires for dual wheels.
(4) Single tires for dual wheels.

43. Instruments. The type of instrument now in general use on electric trucks is the ampere-hour meter. This instrument gives a visual indication of the condition of charge of the battery, both while operating the vehicle and while charging, and also facilitates automatic charging.

44. The rating of electric trucks, according to the recommendations of the Society of Automotive Engineers, is as follows:

Speed rating. The electric-vehicle speed rating shall be based on continuous operation with one-half load over hard, smooth and level roads or pavements at the average battery voltage.

Mileage rating. The electric-vehicle mileage rating shall be based on a continuous run at the S. A. E. rated speed with one-half load over hard, smooth and level roads or pavements.

45. Commercial tractors are identical in design with the commercial truck, except as regards the mounting of the battery, shorter wheel base and lower operating speeds. Tractors may be generally subdivided into three classes:

(1) Trailer type.
(2) Semi-trailer type.
(3) Crane type.

46. The trailer type of tractor comprises those equipped with spring draw bars or pintle hooks, designed to haul a trailer or a train of trailers. In this type the battery is generally mounted in a steel compartment on top of the frame side sills. This type is usually capable of hauling a trailing load of ten to twenty tons gross weight.

47. Approximate Specifications of Various Types of Commercial Trucks and Tractors

Load capacity	Make and model	Chassis, wt. in lb. (excluding battery)	Approx. body allowance in lb.	Approximate battery equipment: No. cells	Make	Type	Max. speed, m.p.h.	Approx. max. mileage per charge
750 lb.	Ward, WS2	1,500	500	60	Edison	A4	13	45
				42	I. C. Exide	MV9		
				42	Phila.	WMT9		
1,000 lb.	Com. Truck, BR1	2,000	600	60	Edison	A5	14	60
	Walker, M	2,300	600	42	I. C. Exide	MV11	15	60
				42	Phila.	WMT13		
1,250 lb.	Ward, WA	2,730	800	60	Edison	A5	12	45
1,500 lb.	Com. Truck, BR2A	2,200	800	42	I. C. Exide	MV11	14	60
	Lansden, BG$\frac{3}{4}$	1,700	800	42	Phila.	WMT13	15	50
2,000 lb.	Atlantic, IC	2,770	1,200				12	....
	Com. Truck, BR2	2,400	1,200	60	Edison	A6	14	60
	Lansden, MC1	2,900	1,200	42	I. C. Exide	MV13	12	50
	Walker, K	2,700	1,200	42	Phila.	WMT15	14	60
	Ward, WB	3,430	1,200				10	40
4,000 lb.	Atlantic, 2C	3,590	1,500				11	40
	Com. Truck, BR4	4,000	1,500	60	Edison	A8	12	60
	Lansden, MD2	4,400	1,500	42	I. C. Exide	MV15	11	40
	Walker, L	3,700	1,500	42	Phila.	WMT17	13	60
	Ward, WD	4,500	1,500				8.5	35
7,000 lb.	Atlantic, 3C	5,220	2,000				10	....
	Com. Truck, AK7	5,800	2,000	60	Edison	A10	11	50
	Lansden, ME3$\frac{1}{2}$	5,700	2,000	42	I. C. Exide	MV19	10	45
	Ward, WF	6,600	2,000	42	Phila.	WMT23	7	30
	Couple gear, H	9,000	2,000				10	30
	Walker, P	5,500	2,000				11	50
10,000 lb.	Atlantic, 5C	6,230	2,500				9	....
	Couple gear, H	10,000	2,500	60	Edison	A12	7	30
	Com. Truck, AK10	6,500	2,500	42	I. C. Exide	MV21	10	50
	Lansden, MT5	7,500	2,500	42	Phila.	WMT25	10	40
	Walker, N	6,400	2,500				10	50
	Ward, WH	8,375	2,500				6	26
Haulage tractor	Walker, 3	4,500	....	60	Edison	A12	10	....
Fifth wheel tractor	Walker, 5	6,000	....	42	I. C. Exide	MV25	10	....
Crane tractor	Walker, 7	7,200	....	42	Phila.	WMT29	10	....

47. Approximate Specifications of Various Types of Commercial Trucks and Tractors *(Continued)*

Load capacity	Make and model	Controller speeds, forward	Tires		Std. wheel base, in.	Per cent. wt. on rear tires	Type of drive
			Front	Rear			
750 lb.	Ward, WS2	4	32×3	32× 3	88	60	Worm
1,000 lb.	Com. Truck, BR1	4	36×3½	36× 3½	89¼	60	Com. truck conc. gear
	Walker, M	5	34×3	36× 3½	94	66	Walker balanced
1,250 lb.	Ward, WA	4	32×3	34× 3½	90	60	Worm
1,500 lb.	Com. Truck, BR2A	4	36×3	36× 3½	92	60	Com. truck. conc. gear
	Lansden, BG¾	4	32×4½*	32× 4½*	90	50	Double red. gear
2,000 lb.	Atlantic, 1C	4	34×4	36× 4	103	65	Chain
	Com. Truck, BR2	4	36×3½	36× 4	101	60	Com. truck conc. gear
	Lansden, MC1	4	36×3	36× 3½	108	60	Chain
	Walker, K	5	34×3½	36× 4	101	66	Walker balanced
	Ward, WB	4	34×3½	36× 4	102	60	Worm
4,000 lb.	Atlantic, 2C	4	34×4	36× 3†	115	65	Chain
	Com. Truck, BR4	4	36×4	36× 4†	116	60	Com. truck conc. gear
	Lansden, MD2	4	36×4	36× 3†	120	60	Chain
	Walker, L	5	38×4	38× 6	112	68	Walker balanced
	Ward, WD	4	36×4	36× 7	114	60	Worm
7,000 lb.	Atlantic, 3C	5	36×5	40× 5†	135	65	Chain
	Com. Truck, AK7	4	36×6	36× 4†	122	55	Com. truck. dbl. sp. gear
	Lansden, ME3½	4	36×5	36× 4†	133	60	Chain
	Ward, WF	5	36×5	36× 8	132	70	Worm
	Couple gear, H	......	36×6	36× 8	96	55	Couple gear wheel
	Walker, P	5	36×5	40× 5†	131	71	Walker balanced
10,000 lb.	Atlantic, 5C	5	36×6	40× 5†	144	65	Chain
	Couple gear, H	5	36×6	36× 8	132	75	Couple gear wheel
	Com. Truck, AK10	4	36×7	36× 5†	132	55	Com. truck. dbl. spur gear
	Lansden, MT5	4	36×6	36× 5†	146	60	Chain
	Walker, N	5	36×6	40× 6†	141	74	Walker balanced
	Ward, WH	5	36×7	40×10	144	70	Worm
Haulage tractor	Walker, 3	5	34×4	40× 6	70	......	Walker balanced
Fifth wheel tractor	Walker, 5	5	36×5	40× 6†	120	......	Walker balanced
Crane tractor	Walker, 7	5	36×6	40× 6	87	......	Walker balanced

NOTE.—* Pneumatic Tires. †Dual Tires.

48. The semi-trailer type of tractor comprises those equipped with fifth wheels, trailer attachments, or swivel connections, designed to haul a semi-trailer. The battery is generally underslung amidship, with sometimes a portion of it mounted directly behind the cab on top of the frame side sills. This type is usually designed to haul a semi-trailer of ten tons gross weight, one-third of the total trailer weight being carried by the rear end of the tractor.

49. The crane type of tractor is equipped with an electrically-operated hoist, with either fixed or swinging boom, designed to carry one ton on the hoist hook, or to hoist two tons with tractor stationary and sprags set. In addition, this type is usually equipped with spring drawbar or pintle hook for hauling trailers.

50. Commercial specifications of trucks and tractors. The principal characteristics of various types of commercial trucks and tractors are given in detail in Par. **47**.

INDUSTRIAL TRUCKS AND TRACTORS

51. Systems of trucking. Three standard systems of storage-battery industrial trucking are in use. The choice of industrial trucks or tractors to meet the particular requirements of any industry should be based on a careful analysis for the purpose of determining which system will best meet the existing or contemplated arrangement of buildings and departments and general transportation facilities. The three systems are:

1. Platform hauling, in which the truck is required to wait while being loaded and unloaded by hand, gravity or crane.
2. Electric self-loading and self-unloading, or semi-self-loading and unloading.
   (*a*) Electric lifting platform and tiering trucks, which eliminate loading and unloading by means of false platforms or skids.
   (*b*) Semi-self-loading or unloading trucks such as crane and hopper trucks.
3. Tractor-trailer systems in which the power unit is coupled to and draws one or more unit loads on wheels, either singly or in trains.

52. Platform trucks are in more general use because of the fact that for most installations the commodities are mixed in character. These trucks are manufactured in numerous models or designs, from which the most suitable type may be selected for a particular application. Among the important factors in making a selection are the height, width and length of the truck desired. The proper height of the platform is dependent, in general, upon the character of the material to be loaded. Standard heights of truck-loading platforms are 12, 17, 18, 20, 24 and 33 in. The character of the runways or roads over which the truck will travel may affect the size of the wheels and consequently the height. The higher platforms are to facilitate the loading of baggage, mail, express, or the like, from car floors and the loading of castings and other heavy materials by means of cranes.

53. The battery mounting in low-platform types is over the driving axle at one end, enabling the use of wheels of small diameter at the other end beneath the low loading platform.

54. The battery mounting in high-platform type is beneath the loading platform. The wheels are of the same size, thus providing ample clearance under the truck. This design provides maximum loading space.

55. Width. The question of width is determined by the aisle, doorway and equipment clearances. With trucks which employ four-wheel steering, the question of clearance is of lesser importance because of the greater flexibility in handling. The standard widths of trucks are 30, 37, 40 and 45 in. The wheel treads should be such as to provide ample stability under all conditions of operation.

56. Length. The length of the standard trucks is dependent upon the character of merchandise to be transported, but the wheel base should not be so long as to impair adaptability. In some cases the ability to control the truck from either end will provide the greatest flexibility in operation.

57. Folding platforms for the operators to stand upon and adjustable end gates facilitate the operation of trucks in congested zones, on elevators and in freight cars.

58. Steering and wheel drive. Trucks are designed for two-wheel drive, with two-wheel steering; four-wheel drive, with four-wheel steering; or two-wheel drive with four-wheel steering.

59. The self-loading and self-unloading types of industrial truck equipment comprise the elevating platform, tiering, crane and hopper trucks, the most popular type being perhaps the electric lifting or elevating platform type.

60. The elevating-platform type of truck is fitted with a lifting platform having a carrying capacity of 4,000 lb. The platform proper is raised and lowered by an electric motor and is designed to run under low portable platforms which are electrically lifted by the platform of the truck and set down without necessity of rehandling. The height of the lift is generally 3 to 5 in. with 4,000-lb. load. The lifting mechanism is operated by an independent motor supplied from the storage battery which propels the truck.

61. Worm lift. In this type the lift is effected by means of a steel worm. The load when raised is sustained on a horizontal plane which takes the entire strain off the lifting mechanism.

62. Screw lift with rocking platform. In this type a motor is direct connected to a worm and gear for driving the elevating mechanism. The spider on which this gear is mounted forms a special clutch adjusted to slip at a fixed overload, in such manner that if the load should catch under an obstruction when lifting, the motor with gear and worm will rotate harmlessly. This clutch is threaded at the center and, when revolving, propels a square multi-thread screw in or out, depending on the direction of motor rotation. The entire mechanism is enclosed in an oil-tight malleable case. The extension of the screw shaft is attached flexibly to the platform on which the load is lifted. The platform is mounted on two heavy links at the rear and one at the front end, providing a three-point support. The action of the screw draws or thrusts the platform backward and upward until the links become practically vertical and raise the load 4½ in. Guides attached beneath the platform steady the load when raised.

A travel limit rod is connected with the motor reversing switch and opens the circuit when the platform reaches either the upper or lower limits of travel. Whenever the motor switch is opened, a special brake on the end of the armature shaft quickly stops the mechanism. The principal advantage of this type is the fact that the operator is able to perform all the labor of loading, transfer and unloading without difficulty.

63. Tiering type. The tiering lifting truck is similar to the elevating platform type (Par. **60**) with the exception that the loads may be lifted any distance from 1 in. to 6 ft. or more, thus making the truck a combination electric-lifting platform truck and tiering machine. This type also has a carrying and raising capacity of 4,000 lb. It will mechanically pick up, transport and deposit its load without the necessity of rehandling, and to a certain extent will perform the work of a portable crane.

64. Crane type. A crane truck embodying an electrically-operated hoist and boom crane mounted on a standard electric platform truck is useful for handling and moving heavy materials or machinery which are not within range of the usual travelling cranes. This type extends the service of the traveling-crane or mono-rail system to the yards or to adjacent buildings where the construction and head room restrict the use of an overhead carrier. Crane trucks are built for lifting capacities of 1,000, 2,000, and 3,000 lb. The boom may be raised or lowered to permit landing the load at any point on the platform of the truck. The boom is also equipped with a swivel base so that it may be swung to either side.

65. Hopper trucks for the handling of coal, ashes, sand and similar materials are in general built in two types, comprising the double side dump and the end dump. In some of the end-dump types the hopper can be dumped by a separate motor energized from the truck battery.

66. The tractor-trailer system of inter-plant transportation has proved to be the more economical and adaptable type of transfer unit where heavy drawbar pull is required and where large tonnage is to be handled daily. This system is comprised of a power unit and as many individual trailers without motive power as may be required. Although any type of truck may be used to haul trailers singly or in trains, the battery capacity is usually insufficient for the continuity of service required.

67. The tractor power unit may be kept in operation almost continuously and does not have to remain idle during the loading and unloading operations, as is the case with load-carrying trucks.

68. Use of the truck as a tractor. Any type of truck may be used to draw trailers, but in general the battery capacity is insufficient for continuous service and trucks as a class are not designed with the necessary drawbar pull for this work. The drawbar pull depends directly upon the weight on the driving wheels and the torque capacity of the motor.

69. Drawbar pull. The normal drawbar pull is the maximum rating under which the tractors are designed to operate continuously, assuming average conditions. The maximum drawbar pull is attained just before the driving wheels begin to slip. The total load in tons which can be moved, including weight of trailers, may be calculated on the basis of 50 lb. drawbar pull per ton, assuming trailers with good bearings traveling on level dry brick, concrete or wood floors.

70. Steering and wheel drive. Tractors are equipped with two-wheel drive, with single-wheel, two-wheel or three-wheel steering, depending upon the type.

71. The frame design follows generally-accepted automotive practice very closely. The frames are made of I-beams, heavy channels, or angle-iron sections, depending on the type of truck or tractor. The frame is supported on the axles either by heavy coil springs or by semi-elliptic springs.

72. Wheel construction. Wheels in general are of the solid or cast type, revolving on roller or ball bearings and equipped with either solid pressed-on rubber or fabric tires. The carrying capacity of solid tires recommended in the Standards of the Society of Automotive Engineers is given in Par. **73.**

73. Industrial Truck Tires
(S. A. E. Standard)

Nominal tire sizes		Wheel dimensions	
Nominal diameter (in.)	Width (in.)	Diameter (in.)	Felloe width (in.)
10	$3\frac{1}{2}$	6	$2\frac{3}{4}$
10	5	6	$4\frac{1}{4}$
16	$3\frac{1}{2}$	12	$2\frac{3}{4}$
16	5	12	$4\frac{1}{4}$
20	$3\frac{1}{2}$	16	$2\frac{3}{4}$
20	5	16	$4\frac{1}{4}$
24	$3\frac{1}{2}$	20	$2\frac{3}{4}$
28	$3\frac{1}{2}$	24	$2\frac{3}{4}$

Wheel diameters shall be 4 in. less than the nominal tire diameters.
The height of the finished tire shall be 2 in. for all sizes.

74. Axles and housings. The axles and housings are made of steel and conform to standard automotive design practice.

75. Transmission. The types of transmission in general use include the worm, chain and spur-gear drives.

76. The chain drive consists of a driving pinion on the motor shaft which engages a sprocket wheel through a silent chain. This sprocket, embodying a compact differential, transmits the mechanical energy through a steel countershaft to sprockets at either end, which in turn engage sprocket wheels on the respective driving wheels by means of roller chains.

77. The worm drive consists of a high-grade steel worm, phosphor bronze wormwheel and differential, with all bearings assembled as a unit in a cast-steel housing, each road wheel being driven through a shaft. All power-transmitting parts are in oil-tight housings and operate in a bath of oil.

78. The spur-gear drive in general comprises a double-reduction spur gear, entirely enclosed and operating in oil.

79. Approximate Specifications of Various Types of Industrial Trucks

Make and model	Capacity in pounds	Weight with Edison battery (lb.)	Weight with Lead battery (lb.)	Drive	No. steering wheels	No. driving wheels	Wheel base (in.)	Tire size (in.) Driving wheels	Tire size (in.) Other wheels
Automatic, R	2,000	1,200	1,400	Worm	2	2	36	16 × 3½	16 × 3½
D	4,000	1,700	1,910	Chain	2	2	39	16	16
E	4,000	2,870	2,980	Worm	2 or 4	2	60	16 or 20	16 or 20
E	4,000	2,300	2,425	Chain	2	2	53	16	16
H	4,000	2,380	2,430	Worm	4	2	53	20 × 3½	9 × 5
H	4,000	2,300	2,350	Gear	4	2	51	16 × 3½	9 × 5
Drop frame bag	4,000	3,250	3,360	Worm	4	2	102	27 × 3½	27 × 3½
Straight frame bag	4,000	3,550	3,660	Worm	4	2	96	27 × 3½	27 × 3½
L	4,000	2,800	2,900	Worm	4	2	54	20 × 3½	9 × 5
L	4,000	2,700	2,750	Gear	4	2	51	16 × 3½	9 × 5
Tiering	4,000	3,450	3,600	Worm	4	2	56	20 × 3½	9 × 5
E crane	4,000	4,070	4,180	Worm	2 or 4	2	60	16 or 20	16 or 20
H crane	4,000	2,440	2,590	Gears	4	2	50	19 × 5	16 × 3½
Elwell Parker, WB	4,000	2,500	2,625	Worm	4	2	78	21½ × 3½	15 × 3½
FC	4,000	2,400	2,525	Worm	4	2	53	21½ × 3½	15 × 3½
IE	3,000	1,800	1,900	Worm	4	2	37	15 × 3½	15 × 3½
ID	4,000	2,050	2,100	Worm	4	2	67	15 × 3½	15 × 3½
IY	4,000	2,450	2,575	Worm	4	2	53	20 × 3½	20 × 3½
IX	4,000	2,700	2,825	Worm	4	4	53	20 × 3½	20 × 3½
Straight frame, S.E., bag	4,000	3,200	3,325	Worm	4	2	85	27¼ × 3½	27¼ × 3½
Drop frame, D.G., bag	4,000	3,250	3,375	Worm	4	2	116	27¼ × 3½	27¼ × 3½
C.F. crane	4,000	5,075	5,200	Worm	4	2	78	21½ × 3½	15 × 3½
E.H., self-ldg	4,000	2,825	2,950	Worm	4	2	58	21½ × 3½	15 × 3½
E.G., self-ldg	4,000	2,800	2,925	Worm	4	2	58	21½ × 3½	15 × 3½

79. Approximate Specifications of Various Types of Industrial Trucks *(Continued)*

Make and model	Loading platform			Control		Max. speed, m.p.h.	Max. turning radii		Max. overall dimensions	
	Length (in.)	Width (in.)	Height (in.)	Single or double end	No. speeds		Outside (in.)	Inside (in.)	Length (in.)	Width (in.)
Automatic, R	64	29	20	S	2	5	77	24	83	29
D	61	36	20	S	3	$5\frac{1}{2}$	106	42	92	38
E	98	41	22–24	S	3	7	178–108	119–49	118	41
E	79	41	20	S	3	8	155	93	117	41
H	64	41	12	S	3	6	98	44	104	41
H	63	41	12	S	3	6	113	57	102	41
Drop frame bag	144	44	11	D	3	6			170	44
Straight frame bag	144	44	33	D	3	6			170	44
L	52	24	10	S	3	6	95	48	103	38
L	55	24	10	S	3	6			103	42
Tiering	52	26	10–95	S	3	6	93	40	117	38
E crane	60	41	23–25	S	3	7	178	119	118	51
H crane		41	12	S	3	6	122	51	102	41
Elwell Parker, WB	93	40	17	S	3	$5\frac{3}{4}$	111	53	137	40
FC	56	40	13	S	3	$5\frac{3}{4}$	184	36	100	40
IE	48	30	20	S	3	$6\frac{1}{2}$	96	45	80	30
ID	88	37	18	S	3	$5\frac{1}{2}$	105	53	112	37
IY	83	40	24	S	3	$5\frac{3}{4}$	96	38	107	40
IX	83	40	24	S	3	5	96	38	107	40
Straight frame, S.E., Bag	137	44	33	D	3	6	156	90	164	45
Drop frame, D.G., bag	137	44	11	D	3	6	198	140	165	45
C.F. crane		40	17	S	3	5	111	53	137	40
E.H., self-ldg	54	26	17	S	3	$5\frac{1}{2}$	92	40	105	36
E.G., self-ldg	54	26	11	S	3	$5\frac{1}{2}$	92	40	105	36

NOTE.—The battery, in general, is comprised of 21 cells Edison A6, or 12 cells lead battery, of 196 to 220 amp.-hr. capacity.

80. Approximate Specifications of Various Types of Industrial Tractors

Make and model	Drawbar pull (lb.)		Weight (lb.)		Drive	Battery equipment				Speeds	
						Edison		Lead			
	Rated	Max.	With Edison battery	With lead battery		No. cells	Amp.-hr. capacity	No. cells	Amp.-hr. capacity	No.	Max. m.p.h.
Automatic, T	600	1,600	2,550	2,750	Worm	.....	.....	.....	.....	3	6
D	250	1,000	2,010	2,420	Chain	.....	.....	.....	.....	3	6
M	800	2,000	3,300	3,525	Worm	.....	.....	.....	.....	3	.....
MT	300	1,200	2,500	2,910	Chain	.....	.....	.....	.....	3	$5\frac{1}{2}$
Elwell Parker, TC	400	1,700	3,350	3,600	Worm	42	225	24	196–220	3	5
TA	300	1,300	2,400	2,450	Worm	30	225	16	196–220	3	$7\frac{1}{2}$
TM	400	2,400	4,000	4,250	Worm	42	225–300	24	196–280	3	6
Mercury	600–750	600–750	1,980 2,270	2,180 2,560	Worm	30	225–300	18	220–315	3	$7\frac{1}{2}$
	1,500	1,500	2,450 2,770	2,660 2,950	Worm	42	225–300	24	220–315	3	$7\frac{1}{2}$

Make and model	No. steering wheels	No. driving wheels	Tire sizes (in.)		Wheel base (in.)	Overall dimensions (in.)			Turning radii	
			Driving	Other		Length	Width	Height	Outside (in.)	Inside (in.)
Automatic, T	1	2	16	16	42	72	40	53	122	51
D	2	2	16	16	39	76	38	55	101	54
M	2 or 4	2 or 4	16 or 20	16 or 20	44	85	40	49	136	74
MT	2	2	16	16	39	71	38	49	108	24
Elwell Parker, TC	4	2	15×5	$15\times3\frac{1}{2}$	38	88	39	52	75	25
TA	1	2	$20\times3\frac{1}{2}$	$15\times3\frac{1}{2}$	40	70	41	52	64	7
TM	2	2	20×5	$20\times3\frac{1}{2}$	45	87	41	52	68	18
Mercury	1	2	$20\times3\frac{1}{2}$	$15\times3\frac{1}{2}$	39	68	39	50	57	.....
	2	2	$20\times3\frac{1}{2}$	$15\times3\frac{1}{2}$	39	68	39	50	71	.....

81. Motors are usually of the 24-volt or 48-volt classification. In worm-drive types the motor is in general mounted so that it can be connected directly to the shaft of the worm. In chain-drive types, and in some worm-drive types, the motor is supported from the chassis frame. See Par. **87** to **107** for details.

82. The controller is usually mounted either on the front of the truck, where it is accessible and readily coupled to the operating mechanism, or underneath the platform and operated by a series of levers and bell cranks. When installed on the front of the truck it can be mounted with the shaft either horizontal or vertical, as desired. The motion of the controller handle is up and down, with horizontal shaft mounting, and right and left with vertical mounting.

83. The safety switch usually serves two purposes as follows: (1) It is operated by the brake pedal, which is spring set, so that when the operator leaves the truck the circuit of the motor is opened and the brake applied; (2) the safety switch is usually so interlocked with the controller drum that the latter must be returned to the "off" position before the safety switch can be closed or the brake released.

84. Interlocking features. Downward pressure of the operator's foot closes the safety switch (Par. **83**) and releases the brake, after which the controller drum can be rotated to any desired position. If the brake be applied, the safety switch opens and the truck stops. The power cannot again be applied until the controller handle is moved to the "off" position. This is known as the **"off-position interlock."** In some applications it is possible to close the safety switch and release the brake with the controller handle in the first position, that is, power can be applied as soon as the brake is released, to prevent losing control of the truck if it is standing on a grade. This is what is known as the **"first-position interlock."** One objection to the latter type of interlock is that the truck can be left standing with the controller handle in the first position and the brake applied, so that an operator returning to the truck and closing the safety switch and releasing the brake without first returning the controller handle to the "off" position would unintentionally start the truck. With the "off-position interlock," the operator must return the control handle to the "off position," before he can release the brake and close the safety switch, this preventing unintentional starting. The "off-position interlock" permits the truck to start without power on a grade, whereas the "first-position interlock" gives complete control on a grade. However, with the former scheme an additional emergency brake can be provided, which, when released, closes the safety switch and allows the controller handle to be moved to the first position.

85. The brake is mounted on the worm shaft or wheels. Braking is effected by a contacting type of brake shoe lined with woven asbestos, actuated by a spring and released by a foot pedal or the operator's platform. In the case of tractors, the brake is sometimes actuated by a spring controlled by the driver's seat, so that the brake is automatically set whenever the driver leaves his seat.

86. The principal characteristics of various commercial types of trucks and tractors are given in the tables in Par. **79** and **80.**

VEHICLE MOTORS

87. Speed-torque characteristic. The inherent characteristic of the motive power for vehicle propulsion should be an increase in effort with decrease in speed. The series motor has this operating characteristic, since its speed varies as an inverse function of the torque, or turning effort. It likewise has a high value of torque per ampere drawn from the battery, throughout its range of operation.

88. Field coils. At present practically all series motors for vehicle service have four field coils, arranged in two pairs, each pair being connected permanently in series. These pairs are arranged normally for connection in series and in parallel. In some special cases, the motors have six field coils instead of four.

89. Type of speed characteristic. The typical speed characteristics of vehicle motors are as follows: (*a*) Flat speed characteristic, in which the speed does not drop rapidly on overloads; (*b*) steep speed characteristic, in which the speed drops rapidly on overloads. A motor having a flat speed characteristic is known as a saturated motor, while a motor having a steep speed characteristic is known as an unsaturated motor. In general, the saturated motor is chosen for application to passenger cars and the unsaturated motor for application to trucks or tractors.

90. General electrical requirements. Vehicle motors are manufacfactured in various sizes, each of which can be supplied in several electrical ratings. The features of design conform closely to railway motor practice, since the conditions of vehicle service are very severe and quite similar to traction service. The motors are designed with liberal commutator capacity and the best possible commutating characteristics, in order to withstand the extreme overloads encountered. In addition, they should possess high efficiency, ruggedness, durability, and general reliability. They must be capable of operating over a considerable range of voltage.

91. Mechanical features. The magnet frame and the commutator end housing are in general cast integral, the outer surface of the cylindrical frame

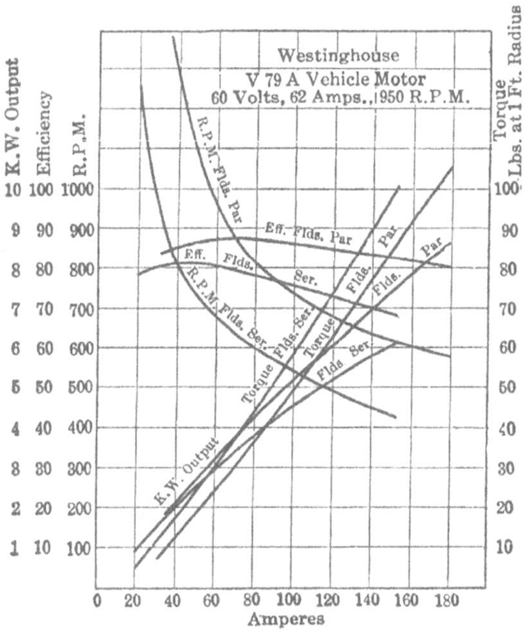

FIG. 8.

usually being machined from end to end. This design lends itself to the application of various forms of suspension brackets. In some designs the magnet frame is made from a rolled steel plate of cylindrical construction having cast-iron, steel, or malleable-iron end housings bolted to each end. The bearing heads of both the commutator and the pinion ends are made of cast iron or bronze and are of the closed type. The armature bearings are of the annular ball type, firmly fastened to the shaft by a retaining nut and protected from dust by a bearing cap. This cap also acts as a retainer for the lubricant. The shaft extension is usually tapered and extends from the pinion end of the motor. The commutators are made of hard-drawn copper with the mica undercut to a depth of approximately 1/16 in.

The armatures after being wound are baked, dipped in a special dipping varnish and baked a second time. The field coils are impregnated with asphaltum under a vacuum process and after being baked are given a final treatment in dipping varnish. This renders both the armature and field windings practically impervious to moisture. The field coils are arranged in two groups, so that they may be connected either in series or in series parallel. Six leads are brought out from the frame, comprising two armature and four field leads, to facilitate connections for the field-control. Two brush arms are in general employed, in order to provide greater accessibility to the brushes. In some designs a shorter overall frame is required, in which case four brush arms are provided, resulting in a more compact design.

92. General motor characteristics. The characteristic curves of torque, speed and current for several types of motors are shown in Figs. 8, 9, 10, 11, and 12, being representative data of two leading makes.

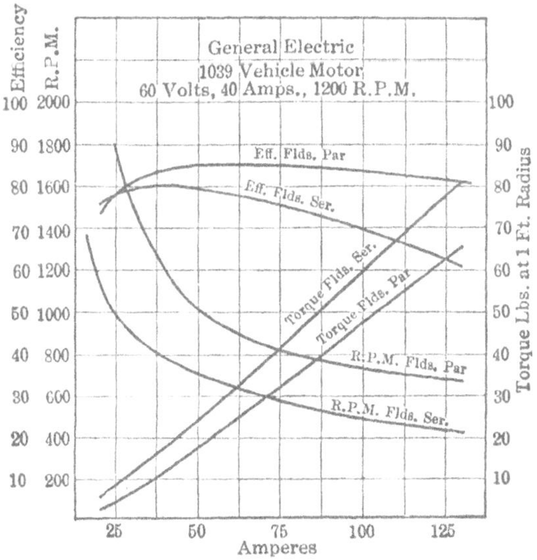

FIG. 9.

93. The performance specification covering vehicle motors may be outlined as follows:

Rating. The normal full-load rating of each motor is based on the specified voltage and amperes. Where series field-coils are arranged for series and parallel connection, the rating is given with field coils connected in parallel.

Speed. The rated speed is based upon the speed of the motor at the end of the specified full-load temperature run. Speeds of individual motors may vary 5 per cent. above or below the rated speed at full load. Approximate speed for loads other than full or rated load are shown on the specified curve.

Efficiency. The efficiencies shown on the specified curve are to be calculated from I^2R losses in the brush contacts and the armature and field windings, based on the measured resistance of these windings at a temperature of 75 deg. cent., and the measured core loss and friction losses, including brush and bearing friction and windage. These losses are to be determined separately for each load at which efficiency is desired. The efficiencies shown on the specified curve are approximate and represent the average of the manufactured product.

Commutation. Each motor shall operate at rated full load with no sparking or burning of the brushes and without blackening of the commutator. Each motor shall operate at any load encountered in normal operation without injurious sparking.

Temperature. After a continuous run (for 4 hours) at the specified voltage and current, no part of the windings shall exceed a temperature of 65 deg. cent. above the surrounding air. Test shall be made on the stand with motor covers on. Temperature shall be measured by thermometer.

94. S. A. E. motor standards. The practice recommended by the Society of Automotive Engineers in regard to electric vehicle motors is as follows:

Motor voltage. Two classes of motors shall be used for electric vehicles (not including industrial trucks and tractors), the first class designed for 80-volt to 85-volt operation, characteristic curves to be furnished by the manufacturers for each voltage. The second class of motors shall be designed for 60-volt to 66-volt operation, the characteristic curves for each of these voltages to be furnished by the manufacturers.

Motor nameplates. Each motor shall carry a nameplate setting forth: Manufacturer's name and address; whether series, shunt, or compound wound; frame size; volts, amperes, and revolutions per minute, for both the high and the low limits of voltage, given in separate figures.

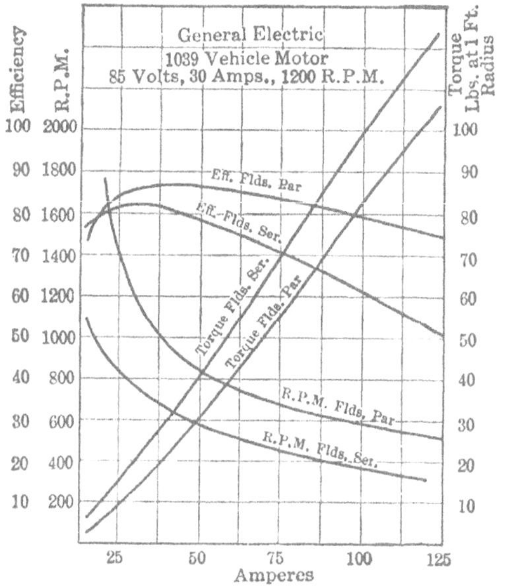

FIG. 10.

95. Motor windings. As the result of the standardization of 40 to 44 cells of lead battery and 60 cells of Edison battery, the motor manufacturers have designed their respective lines of motors so that either of two windings can be provided for each frame. The one, an 80-volt to 85-volt rating, is designed for use with a 40-cell to 44-cell lead battery, and the other, a 60-volt rating, is designed for use with a 60-cell Edison battery. The two windings are so designed that the motor will have the same speed and torque at normal rated load for both voltage ratings. This can be seen by comparison of the curves in Figs. 9 and 10 and Figs. 11 and 12.

96. Standard motor sizes. The standard motors of two well-known manufacturers are listed in the table in Par. **98,** showing the ratings and the weights.

97. Choice of number of motors. The choice of one, two, or four motors is dependent upon the type of transmission employed. The single-motor drive is generally favored because of its high efficiency, light weight, and simplicity of control. The two-motor drive gives slightly better performance when the stops per mile are very frequent, due to the elimination of running resistance in the first notches of the controller, but it is doubtful whether this advantage offsets the additional weight and maintenance necessitated by the additional motor.

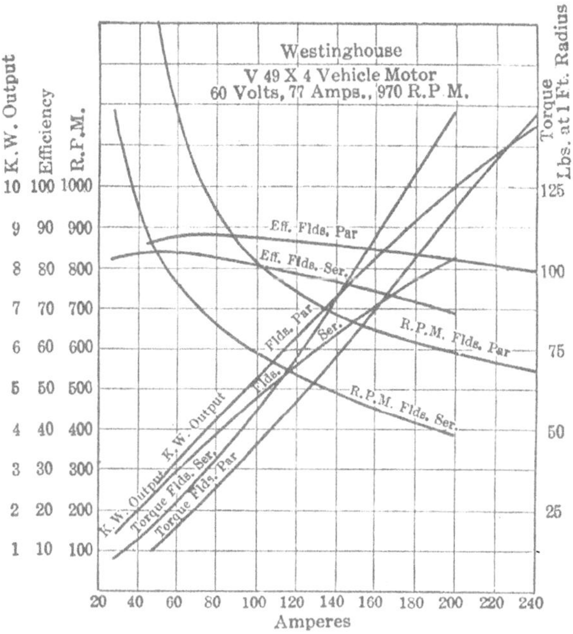
Westinghouse
V 49 X 4 Vehicle Motor
60 Volts, 77 Amps., 970 R.P.M.
K.W. Output
Efficiency
R.P.M.
Torque
Lbs. at 1 Ft. Radius
Eff. Flds. Par
Eff. Flds. Ser.
R.P.M. Flds. Par
R.P.M. Flds. Ser.
K.W. Output Flds. Par
K.W. Output Flds. Ser.
Torque Flds. Ser.
Torque Flds. Par
Amperes

Fig. 11.

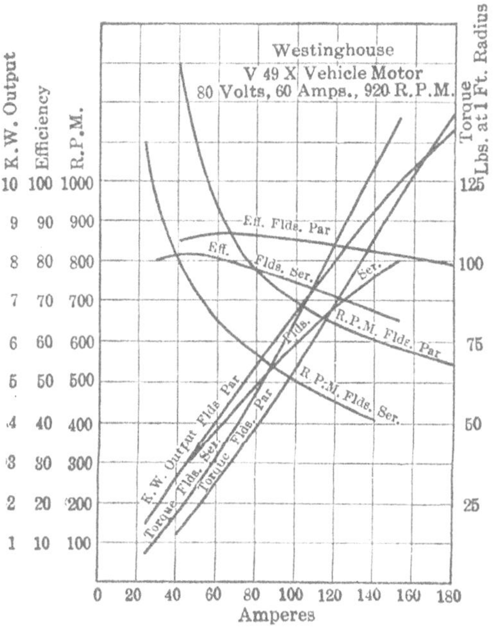
Westinghouse
V 49 X Vehicle Motor
80 Volts, 60 Amps., 920 R.P.M.
K.W. Output
Efficiency
R.P.M.
Torque
Lbs. at 1 Ft. Radius
Eff. Flds. Par
Eff. Flds. Ser.
R.P.M. Flds. Par
R.P.M. Flds. Ser.
K.W. Output Flds. Par
Flds. Ser.
Torque Flds. Ser.
Torque Flds. Par
Amperes

Fig. 12.

98. Standard Vehicle Motor Ratings

(General Electric Co.)

Type	Volts	Amp.	R.p.m.	Wt. (lb.)	Type	Volts	Amp.	R.p.m.	Wt. (lb.)
GE 1086	24	32	1,900	45	GE 1048	60	40	1,600	265
	36	21	1,800	45		85	30	1,600	265
GE 1083	24	40	1,500	75	GE 1064	48	50	1,500	315
GE 1065	60	17	1,600	130	GE 1039	60	40	1,200	310
	85	12	1,600	130		85	30	1,200	310
GE 1090	24	45	1,200	170	GE 1022	60	60	1,200	425
	24	50	1,500	170		85	40	1,200	425
	32	35	1,500	170	GE 1030	60	70	850	530
	40	35	2,000	170		85	50	850	530
GE 1091	24	65	1,700	200	GE 1027	60	85	900	660
	32	60	1,500	200		85	60	900	660
	48	45	1,700	200	GE 1087	80	120	900	890

(Westinghouse Elec. & Mfg. Co.)

Typr	Volts	Amp.	R.p.m.	Wt. (lb.)	Type	Volts	Amp.	R.p.m.	Wt. (lb.)
V-62-X-4	60	36	1,160	145	V-50-X-4	60	57	1,190	300
V-62-X	80	28	1,160	145	V-50-X	80	43	1,190	300
V-51-X-4	60	44	1,100	190	V-49-X-4	60	77	970	430
V-54-X	80	33	1,090	190	V-49-X	80	60	920	430
V-77-A	80	40	950	250	V-43-X-4	60	120	800	600
V-79-A	60	62	910	280	V-43-X	80	90	840	600

99. Choice of proper size of motor. In applying a motor to a passenger car or truck, it has been found by experience that a motor having a normal current rating equal to the current required to propel the car on hard level, smooth pavement at its rated speed and load capacity will, in general, have sufficient capacity to propel the car under general operating conditions without exceeding the safe temperature limits. The following general expressions can be used to determine the capacity of motor required, or other variable involved:

$$TE = W \times R \text{ (lb.)} \tag{1}$$

$$EI = \frac{TE \times MPH}{0.502 \times e_g \times e_m} \text{ (watts)} \tag{2}$$

$$I = \frac{TE \times MPH}{0.502 \times E \times e_g \times e_m} \text{ (amp.)} \tag{3}$$

Substituting the value of TE from equation (1),

$$I = \frac{W \times R \times MPH}{0.502 \times E \times e_g \times e_m} \text{ (amp.)} \tag{4}$$

also

$$I = \frac{W \times w.h. \times MPH}{E} \text{ (amp.)} \tag{5}$$

$$MPH = \frac{N \times D}{336 \times g.r.} \tag{6}$$

$$T = \frac{TE \times D}{g.r. \times 24 \times e_g} \tag{7}$$

$$= \frac{W \times R \times D}{g.r. \times 24 \times e_g} \tag{8}$$

Where TE = tractive effort required at rims of driving wheels, in lb.
W = gross weight of vehicle in tons.
R = total frictional resistance, in lb. per ton.
E = voltage at motor terminals.
I = current in amperes.
MPH = speed of vehicle in miles per hour.
T = motor torque in lb. at 1 ft. radius.
e_g = efficiency of gearing, assumed usually as 95 per cent. for single reduction and 90 per cent. for double reduction.
e_m = efficiency of motor, usually 85 to 88 per cent., average 86 per cent.
$w.h.$ = watt hours per ton mile
N = speed of motor armature in revolutions per minute.
D = diameter of driving wheels in inches.
$g.r.$ = total speed reduction ratio between motor shaft and driving wheels.

100. The rolling resistance of a wheel is due theoretically to the yielding or indentation of the road, which causes the wheel continually to mount a slight incline. It is also affected by the yielding or deformation of the wheel tread at the point of contact with the road. The resistance is measured by the horizontal force necessary at the axle to lift it over the obstacle or roll it up the incline surface. The rolling resistance varies with the diameter of the wheel, the width and composition of the tire, the construction of the tire, the method of attaching the tire to the wheel rim, the speed, the nature of the road surface, the grade and the spring suspension.

101. The air resistance varies approximately as the square of the speed, but is usually not considered in electric vehicle practice because the speeds are relatively so low that it is negligible.

102. Vehicle resistance is given by the best authorities as having the following values:

Steel-tired wheels on steel rails..........	6 to 20 lb. per ton
Pneumatic tired wheels on level hard asphalt pavement....................	15 to 35 lb. per ton
Solid-tired wheels on level hard asphalt pavement........................	20 to 40 lb. per ton
Steel-tired wheels on average roads, approximately.....................	50 lb. per ton

103. The relative values of vehicle resistance on various level road surfaces, as compared to the value on hard level asphalt, are given in the following table.

Road surface	Relative value of vehicle resistance	Road surface	Relative value of vehicle resistance
Asphalt, hard.......	1.00	Snow packed............	1.3
Macadam..........	1.15	Snow fairly hard, without grips................	1.7
Tarvia.............	1.14	Snow fairly hard, with grips................	1.9
Glazed brick........	1.35	Snow, soft..............	2.1
Wood block.........	1.1		
Granite block.......	2.0		

104. The vehicle resistance for a passenger car may be assumed as 20 to 30 lb. per ton for operation on level hard smooth asphalt.

105. The vehicle resistance for electric trucks on level hard smooth asphalt has been found by experience to be approximately as follows:

Capacity of truck, lb.	Pounds per ton	Watthours per ton mile
1,000	40	90
2,000	40	85
4,000	35	80
7,000	30	75
10,000	30	75

106. Effect of grades. The effect of grades is to increase or decrease the value of the vehicle resistance by 20 lb. per ton for each per cent grade.

107. The energy consumption and power demand of electric vehicles are governed by the gross weight, speed, character of roads, grades and frequency of stops.

VEHICLE CONTROL

108. Two general methods. There are two general methods of control. In the first the battery cells are connected permanently in series and in the other the battery is commutated, that is, the battery cells are divided into two equal groups, which can be connected in parallel or series combination.

109. In the first method (Par. **108**) two resistance notches are generally used before the "field-series" notch is reached. Between this position and the "fields-parallel" position an intermediate position is provided in which a resistance is in parallel circuit or shunt with the field coils. This is not absolutely necessary but it is done so that the increase of current or torque from notch to notch is uniform. In some cases another position is provided beyond the normal running or "field-parallel" position in which a resistance is in parallel with the field coils to secure a still higher operating speed. This is what is commonly known as an emergency speed and is usually found only on controllers designed principally for application to passenger cars. The controller is often arranged so that it is necessary for the operator to maintain the handle in this position against the force of a spring tending to turn the handle back to the series-parallel position. A controller of this first general type provides two economical running speeds, one with "fields series" and the other with "fields parallel."

110. The second method affords a slightly better efficiency of operation in that energy is not dissipated through ohmic resistance during acceleration. Formerly this method was highly objectionable because the controllers were not of the continuous-torque type, that is, the circuit was interrupted when the drum was rotated from notch to notch. Furthermore some trouble was encountered because of unequal discharge of the two sections of the battery. Continuous-torque controllers have been developed, which in addition to the foregoing feature have an improved system of battery bridging that obviates the trouble arising from the unbalancing of the two sections of the battery. The bridging is accomplished in the momentary transition from "batteries parallel, field-series," to the position "battery series, field series" by consecutive momentary transition combinations as follows:

(*a*) A resistance is in parallel circuit with a battery section in parallel, one group of field circuits open circuited.

(*b*) Battery sections in series, resistance connected between the two sections, fields in series, one group short circuited.

(*c*) Battery sections in series, resistance connected between the two sections, field coils in series.

It will be seen that in the transition both sections of the battery are discharged at equal rates, never becoming unbalanced. With this second method, four economic running speeds are provided as compared with two for the first, these being "fields series" and "fields parallel" connections, with the battery sections connected in parallel and in series respectively. In the operation of a vehicle where there are numerous acceleration periods, this method results in more economical operation. It also permits of economical operation at low speeds in congested traffic zones.

111. General features. In general controllers should be so designed and built that compactness, ruggedness, ease of operation and accessibility are secured.

112. Drum type. Controllers are usually of the drum type, employing one drum for manipulating the connections for both forward and reverse speeds, or two drums, comprising a main drum for controlling the speed combinations and a reverse drum for reversing the armature circuit. In the latter type the main drum and the reverse drum are interlocked, so that the reverse drum is inoperative while the main drum is in any "on" position, it being necessary to return the main drum to the "off" position before the reverse drum can be operated.

113. Controllers for passenger cars usually have a greater number of speeds forward than those for application to electric trucks, since the operating speeds of the passenger car are higher and uniform acceleration is necessary. In some controllers for passenger car service an emergency speed is obtained by shunting a resistance around the parallel fields of the motor (Par **109**).

114. The various speed notches on the controller should be indicated positively by action of the star wheel and pawl. In the case of a single-drum type of controller, there should be a considerable break between the first forward and the first reverse positions. In this type the reversing operation is usually effected at the "off" position by a sidewise motion of the handle against the positive pressure of a spring, permitting the handle to be operated in a backward direction from the "off" position in an off-set slot. This arrangement prevents the sudden reversal of the motor by the careless operation of the controller handle by the driver.

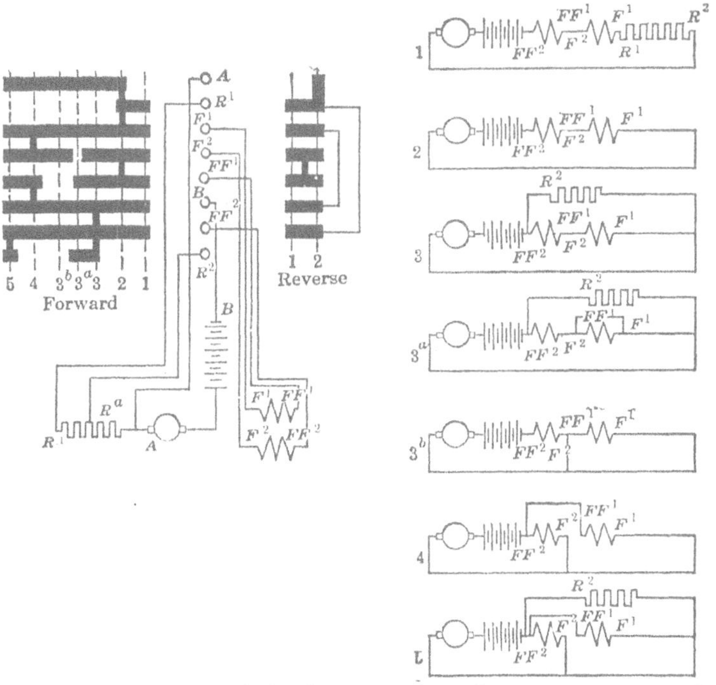

FIG. 13.—Connections and development of continuous torque control for single motor equipment.

115. Angular travel. The angular throw of the handle should be as small as possible, but the peripheral travel of the contact surfaces of the control drum must be great enough to rupture the arcs when the drum is rotated from notch to notch.

116. Drum construction. The drum bases or cylinders are made of wood or of molded material, attached by spiders to the operating shaft. The main drums are operated either by a handle attached directly to the shaft, or through a pinion and sector gear, the latter being fastened to a stub shaft, to which the operating handle is attached. With the latter form of operating mechanism the gear reduction is approximately 3 to 1 so that the distance through which the handle moves is grealty reduced in comparison with that through which the controller drum rotates in moving from notch to notch.

117. Continuous torque operation is a desirable feature of control because it permits the current to flow uninterruptedly while the controller drum is being rotated from notch to notch, forward or reverse, thereby securing maximum life of contact fingers and plates and smooth efficient acceleration. In the system of control ordinarily used, the interruption of the current will take place only in the transition from the series to the series-parallel field connection.

118. Typical controller connections embodying the continuous-torque feature are illustrated in Fig. 13. The continuous torque feature is effected as follows: On the third position the motor fields are in series and shunted with the resistance R^2. As the controller drum is rotated between the positions 3 and 4, a short-circuit is introduced at 3^a around one set of fields. Referring to the drum development, it will be noted that this is caused by the contact finger which bridges across the gap on the contact F^2, thus momentarily connecting F^1 and FF^1. In the next momentary transition 3^b, the shunt R^2 is removed, and the field connection FF^1 is opened; the

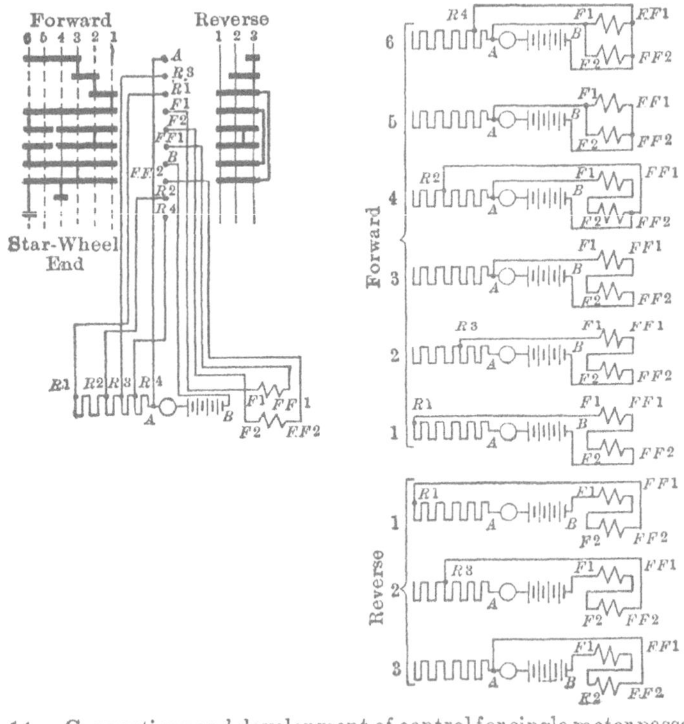

FIG. 14.—Connections and development of control for single motor passenger car equipment.

contact finger F^2 has now passed across the gap, leaving the field terminal F^2 connected to F^1. In the next position, which is the fourth, the terminal FF^1 is connected to FF^2, placing the fields in parallel. During the transition only one set of fields is used. However, with a 4-pole motor having diametrically opposite coils connected in series in each group, it has been found that there is very little momentary increase in operating current.

119. Single-motor control. The controller connections shown in Fig. 13 are representative of those for single-motor trucks, while those in Fig. 14 are representative of single-motor passenger cars.

120. Controllers for industrial trucks and tractors are of the drum type and are usually arranged for three speeds forward and reverse. The first speed is obtained by connecting the two sections of the motor field in series with a resistance, the second speed by short-circuiting the resistor and the third speed by paralleling the two sections of the motor field. The principal function of the three speed points is to start the truck gently, as very little actual speed control is required. In tractor equipments the resistor point is useful, however, because slow speed is desired when coupling trailers.

121. Double-motor control. With double-motor equipments the general scheme of control is the same as for single-motor equipments, a series-parallel commutation of the motors being usually employed, with series resistance for starting and intermediate steps. At starting, the motors are connected in series with all resistance in circuit.

122. The control resistors are usually either of cast iron or cast alloy. In some cases, to prevent rusting, the grids are copper plated and treated with aluminum paint. The grids are usually assembled on tie rods, mica insulated and held between two pressed or punched steel end-frames.

123. A main switch is usually provided, so that the controller can be isolated from the battery circuit when the battery is on charge.

STORAGE BATTERIES

124. General. The subject of storage batteries as a whole is treated in Section 20. The following discussion relates exclusively to storage batteries in vehicle service and is intended to supplement the subject matter of Section 20.

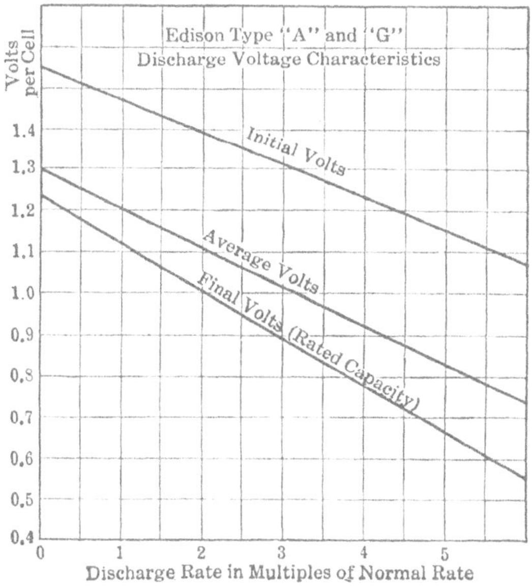

FIG. 15.

125. Three types of batteries in general use on electric vehicles are the Edison, the Iron Clad Exide and the Philadelphia Diamond Grid types *WMT* and *WTXI*. The Philadelphia type *WTXI* is of the thin-plate type and recommended only for passenger car service.

126. Principal characteristics of the Edison battery are embodied in the table in Par. **129,** the Iron Clad Exide in Par. **130,** the Philadelphia type *WMT* in Par. **131** and the Philadelphia type *WTXI* in Par. **132.**

127. The battery discharge in vehicle service, in all applications, occurs at varying rates and is accompanied by intermittent periods of rest. The total elapsed time of discharge is ordinarily eight hours or more. The reduction in capacity due to high discharge rates largely disappears when the rates are intermittent. Therefore, the normal discharge-rate capacity specified by the manufacturers will be available, provided the total elapsed time during which the discharge takes place is not less than 6 hr.

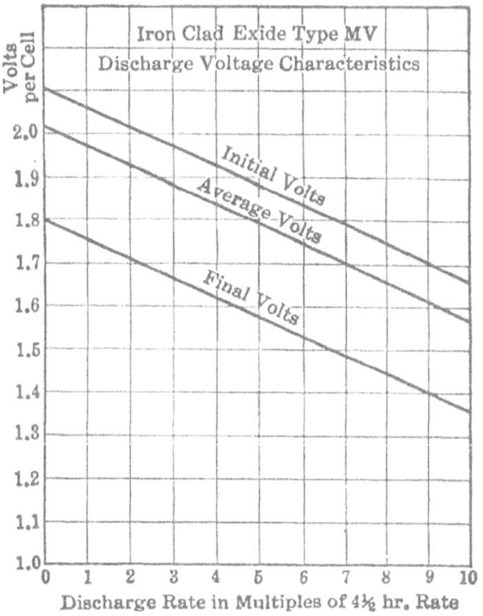

FIG. 16.

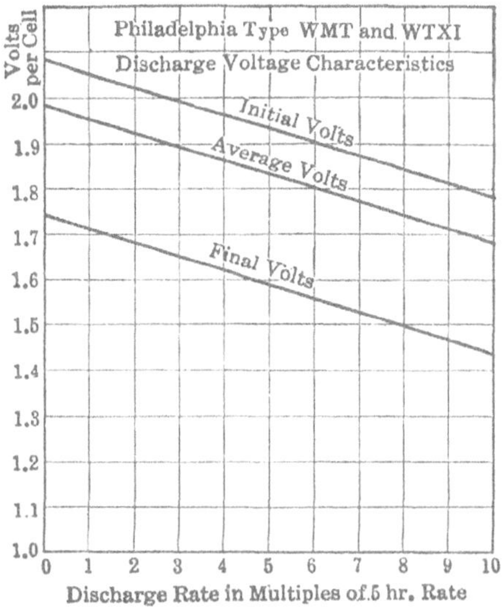

FIG. 17.

128. The discharge voltage characteristics at various multiples of the normal discharge rate are given in Fig. 15 for the Edison battery, Fig. 16 for the Iron Clad Exide Battery and Fig. 17 for the Philadelphia types *WMT* and *WTXI*.

129. Data on Edison Type "A" and "G" Cells

(Applying to Individual Cells)

Type of cell	A4	A5	A6	A8	A10	A12	G4	G6	G7	G9	G11	G14	G18
Amp.-hr. capacity (rated)	150	187.5	225	300	375	450	100	150	175	225	275	350	450
Kw.-hr. capacity (rated)	0.180	0.225	0.270	0.360	0.450	0.540	0.120	0.180	0.210	0.270	0.330	0.420	0.540
Discharge rate:													
(a) Normal, amp.	30	37.5	45	60	75	90	30	45	52.5	67.5	82.5	105	135
(b) Max. for intermittent discharge, amp.	180	225	270	360	450	540	180	270	315	405	495	630	810
Normal charging rate, amp.	30	37.5	45	60	75	90	30	45	52.5	67.5	82.5	105	135
Dimensions, bottomless trays (in.):													
(a) Width	$6\frac{1}{8}$	$6\frac{1}{8}$	$6\frac{1}{8}$	$6\frac{1}{8}$	$7\frac{3}{4}$	9	$6\frac{1}{8}$	$6\frac{1}{8}$	$6\frac{1}{8}$	$5\frac{1}{8}$	$6\frac{1}{8}$	$7\frac{3}{4}$	$9\frac{5}{8}$
(b) Height overall—filler cap closed	$14\frac{3}{8}$	$14\frac{3}{8}$	$14\frac{3}{8}$	$14\frac{3}{4}$	$14\frac{3}{4}$	$15\frac{1}{2}$	$14\frac{3}{8}$	$14\frac{3}{8}$	$14\frac{3}{8}$	$14\frac{3}{8}$	$14\frac{3}{4}$	$14\frac{3}{4}$	$15\frac{1}{2}$
Height overall—filler cap open	16	16	16	$16\frac{3}{8}$	$16\frac{3}{8}$	17	16	16	16	16	$16\frac{3}{8}$	$16\frac{3}{8}$	17
(c) Length: 1-cell tray	$4\frac{5}{8}$	$5\frac{1}{4}$	$5\frac{3}{4}$	$7\frac{1}{4}$	$7\frac{1}{2}$	$7\frac{1}{2}$	$4\frac{1}{8}$	$4\frac{3}{4}$	$5\frac{1}{4}$	5	$7\frac{1}{4}$	$7\frac{1}{2}$	$7\frac{1}{2}$
2-cell tray	$7\frac{3}{4}$	9	10	$12\frac{7}{8}$	$13\frac{1}{2}$	$13\frac{1}{2}$	$6\frac{3}{4}$	8	9	$10\frac{3}{8}$	$12\frac{7}{8}$	$13\frac{1}{2}$	$13\frac{1}{2}$
3-cell tray	$10\frac{7}{8}$	$12\frac{3}{4}$	$14\frac{1}{4}$	$18\frac{3}{8}$	$19\frac{1}{2}$	$19\frac{1}{2}$	$9\frac{3}{8}$	$11\frac{1}{4}$	$12\frac{3}{4}$	$14\frac{7}{8}$	$18\frac{3}{8}$	$19\frac{1}{2}$	$19\frac{1}{2}$
4-cell tray	14	$16\frac{3}{8}$	$18\frac{1}{2}$	$24\frac{1}{8}$	$25\frac{7}{8}$	$25\frac{7}{8}$	$12\frac{3}{8}$	$14\frac{7}{8}$	$16\frac{3}{8}$	$19\frac{5}{8}$	$24\frac{1}{8}$	$25\frac{7}{8}$	$25\frac{7}{8}$
5-cell tray	$17\frac{1}{8}$	20	$22\frac{3}{4}$	$29\frac{3}{4}$	$32\frac{1}{4}$	$32\frac{1}{4}$	15	$18\frac{1}{8}$	20	$24\frac{1}{8}$	$29\frac{3}{4}$	$32\frac{1}{4}$	$32\frac{1}{4}$
6-cell tray	$20\frac{1}{4}$	$23\frac{5}{8}$	$27\frac{3}{4}$	36	......	......	$17\frac{5}{8}$	$21\frac{3}{8}$	$23\frac{5}{8}$	$28\frac{1}{2}$	36	......	......
7-cell tray	$23\frac{1}{4}$	$28\frac{1}{2}$	32	$41\frac{5}{8}$	......	......	$20\frac{5}{8}$	25	$28\frac{1}{2}$	$33\frac{3}{8}$	$41\frac{5}{8}$	......	......
8-cell tray	$27\frac{1}{4}$	$32\frac{1}{4}$	$36\frac{1}{4}$	$47\frac{1}{8}$	......	......	$23\frac{1}{4}$	$28\frac{1}{4}$	$32\frac{1}{4}$	$37\frac{3}{4}$	$47\frac{1}{8}$	......	......
9-cell tray	$30\frac{3}{8}$	36	$41\frac{1}{8}$	$53\frac{5}{8}$	......	......	$25\frac{7}{8}$	$31\frac{1}{2}$	36	$42\frac{1}{4}$	$53\frac{5}{8}$	......	......
10-cell tray	$33\frac{1}{4}$	$40\frac{1}{8}$	$45\frac{1}{8}$	......	......	......	$28\frac{7}{8}$	$35\frac{1}{8}$	$40\frac{1}{8}$	47	......	......	......
11-cell tray	$36\frac{3}{8}$	$43\frac{7}{8}$	......	......	......	......	$31\frac{1}{2}$	$38\frac{3}{8}$	$43\frac{7}{8}$	......	......	......	......
12-cell tray	$40\frac{1}{8}$	......	......	......	......	......	$34\frac{1}{8}$	$41\frac{5}{8}$	......	......	......	......	......
Weight in lb.:													
(a) Solution	3.1	3.6	4.1	6.1	7.7	9.4	2.4	3.4	4.0	4.7	5.8	7.3	10.2
(b) Complete cell	13.8	16.7	19.4	26.8	33.3	40.6	10.9	15.1	17.4	21.1	27.3	34.0	44.5
(c) Complete cell, including tray	14.9	17.5	20.8	29.0	36.2	43.9	13.0	16.8	18.2	23.0	29.6	36.9	47.8

130. Data on Iron-clad Exide Type MV Cells

(Applying to Individual Cells)

Number of plates	7	9	11	13	15	17	19	21	23	25	27	29	31	33
Amp.-hr. capacity (6-hr. rate)	102	136	170	204	238	272	306	340	374	408	442	476	510	544
Kw.-hr. capacity (6-hr. rate)	0.201	0.268	0.335	0.402	0.469	0.536	0.603	0.670	0.737	0.804	0.871	0.938	1.005	1.072
Discharge rate ($4\frac{1}{2}$-hr. rate), amp.	21	28	35	42	49	56	63	70	77	84	91	98	105	112
Normal charging rates:														
(*a*) Starting, amp.	19	24	30	35	40	45	51	56	61	66	71	76	81	86
(*b*) Finishing, amp.	8	10	12	14	16	18	20	22	24	26	28	30	32	34
Cell dimensions (not including tray):														
(*a*) Outside length (in.)	$2\frac{13}{16}$	$3\frac{9}{16}$	$4\frac{5}{16}$	$5\frac{1}{16}$	$5\frac{7}{8}$	$6\frac{5}{8}$	$7\frac{3}{8}$	$8\frac{1}{8}$	$8\frac{7}{8}$	$9\frac{5}{8}$	$10\frac{3}{8}$	$11\frac{1}{8}$	$11\frac{7}{8}$	$12\frac{5}{8}$
(*b*) Outside width (in.)	$6\frac{3}{16}$	$6\frac{3}{16}$	$6\frac{3}{16}$	$6\frac{3}{16}$	$6\frac{1}{4}$	$6\frac{1}{4}$	$6\frac{1}{4}$	$6\frac{1}{4}$	$6\frac{1}{4}$	$6\frac{1}{4}$	$6\frac{1}{4}$	$6\frac{1}{4}$	$6\frac{1}{4}$	$6\frac{1}{4}$
(*c*) Overall height (in.)	15	15	15	15	$15\frac{1}{16}$	$15\frac{1}{16}$	$15\frac{1}{16}$	$15\frac{1}{16}$	$15\frac{1}{16}$	$15\frac{1}{16}$	$15\frac{1}{16}$	$15\frac{1}{16}$	$15\frac{1}{16}$	$15\frac{1}{16}$
Weight in pounds:														
(*a*) Electrolyte	$4\frac{1}{4}$	$5\frac{1}{2}$	$6\frac{3}{4}$	8	$9\frac{1}{4}$	$10\frac{1}{2}$	$11\frac{3}{4}$	13	$14\frac{1}{4}$	$15\frac{1}{2}$	$16\frac{3}{4}$	18	$19\frac{1}{4}$	$20\frac{1}{2}$
(*b*) Complete cell	$23\frac{3}{4}$	30	$36\frac{1}{2}$	$42\frac{3}{4}$	$49\frac{1}{2}$	$57\frac{1}{4}$	$63\frac{1}{2}$	70	$76\frac{1}{4}$	$82\frac{3}{4}$	89	$95\frac{1}{4}$	$101\frac{3}{4}$	108
(*c*) Complete cell, including tray	27	35	42	49	57	66	73	81	88	95	102	110	117	124

131. Data on Philadelphia Type WMT Cells
(Applying to Individual Cells)

Number of plates	9	11	13	15	17	19	21	23	25	27	29	33
Amp.-hr. capacity (6-hr. rate)	124	155	186	217	248	279	310	341	372	403	434	496
Kw.-hr. capacity (6-hr. rate)	0.244	0.305	0.366	0.428	0.489	0.550	0.611	0.672	0.733	0.794	0.855	0.977
Discharge rates:												
(a) 6-hr. rate, amp.	20.7	25.8	31	36.2	41.3	46.5	51.7	56.8	62	67.2	72.3	82.7
(b) 5-hr. rate, amp.	24	30	36	42	48	54	60	66	72	78	84	96
Normal charging rates:												
(a) Starting, amp.	24	30	36	42	48	54	60	66	72	78	84	96
(b) Finishing, amp.	6	7.5	9	10.5	12	13.5	15	16.5	18	19.5	21	24
Cell* dimensions (not including tray):												
(a) Outside length (in.)	$3\frac{1}{16}$	$3\frac{9}{16}$	$4\frac{5}{16}$	$4\frac{7}{8}$	$5\frac{9}{16}$	$6\frac{1}{4}$	$6\frac{7}{8}$	$7\frac{1}{2}$	$8\frac{1}{8}$	$8\frac{3}{4}$	$9\frac{3}{8}$	$10\frac{5}{8}$
(b) Outside width (in.)	$6\frac{3}{16}$	$6\frac{3}{16}$	$6\frac{3}{16}$	$6\frac{3}{16}$	$6\frac{3}{16}$	$6\frac{1}{4}$	$6\frac{1}{4}$	$6\frac{1}{4}$	$6\frac{1}{4}$	$6\frac{1}{4}$	$6\frac{1}{4}$	$6\frac{1}{4}$
(c) Overall height (in.)	$14\frac{15}{16}$	$14\frac{15}{16}$	$14\frac{15}{16}$	$14\frac{15}{16}$	$14\frac{15}{16}$	15	15	15	15	15	15	15
Weight in lb.:												
(a) Electrolyte	5.0	6.1	7.1	8.3	9.3	10.5	11.5	12.7	14.0	15.2	16.2	18.6
(b) Complete cell	27.5	32.5	37.5	42.5	47.5	53.0	58.0	63.5	70.0	76.0	82.0	94.0
(c) Complete cell, including tray	30.8	35.8	41.3	46.8	52.3	58.3	63.8	69.9	77.0	83.6	90.2	103.4

* Dimensions embody S.A.E. standard jars having $\frac{5}{32}$-in. walls for 9 to 17 plates inclusive, and $\frac{3}{16}$-in. walls for 19 to 33 plates all with $2\frac{1}{4}$-in. ribs. Height specified is from bottom of jar to highest projecting point at top of cell.

132. Data on Philadelphia Type WTXI Cells
(Applying to Individual Cells)

Number of plates	9	11	13	15	17	19	21
Amp.-hr. capacity:							
(*a*) 6-hr. rate	102	127	153	178	204	229	255
(*b*) 5-hr. rate	98	122	147	171	196	220	245
Kw.-hr. capacity:							
(*a*) 6-hr. rate	0.201	0.251	0.301	0.352	0.402	0.452	0.502
(*b*) 5-hr. rate	0.191	0.238	0.286	0.334	0.382	0.430	0.478
Discharge rates:							
(*a*) 6-hr. rate, amp	17.0	21.2	25.5	29.7	34.0	38.2	42.5
(*b*) 5-hr. rate, amp	19.6	24.5	29.4	34.3	39.2	44.1	49.0
Normal charging rates:							
(*a*) Starting, amp	17	22	25	30	34	38	43
(*b*) Finishing, amp	4	5	6	8	9	10	12
Cell* dimensions (not including tray):							
(*a*) Outside length (in.)	$2\frac{5}{16}$	$2\frac{13}{16}$	$3\frac{5}{16}$	$3\frac{13}{16}$	$4\frac{1}{4}$	$4\frac{13}{16}$	$5\frac{5}{16}$
(*b*) Outside width (in.)	$6\frac{1}{8}$	$6\frac{1}{8}$	$6\frac{1}{8}$	$6\frac{1}{8}$	$6\frac{1}{8}$	$6\frac{1}{8}$	$6\frac{1}{8}$
(*c*) Overall height (in.)	$14\frac{1}{4}$	$14\frac{1}{4}$	$14\frac{1}{4}$	$14\frac{1}{4}$	$14\frac{1}{4}$	$14\frac{1}{4}$	$14\frac{1}{4}$
Weight in lb.:							
(*a*) Electrolyte	3.25	4.25	5.25	6.0	6.75	7.75	8.75
(*b*) Complete cell	17.25	21.25	25.25	29.0	33.0	37.0	41.0
(*c*) Complete cell, including tray	19.0	23.4	27.8	31.9	36.3	40.7	45.1

* Dimensions embody S.A.E. standard jars having $\frac{1}{8}$-in. walls and $2\frac{3}{4}$-in. ribs. Height specified is from bottom of jar to highest projecting point at top of cell.

BATTERY CHARGING

133. General Principle. To charge a battery fully after discharge it is necessary to pass through the cells in a direction opposite to that of discharge an amount of current equivalent in ampere-hours to that taken out on discharge, plus an excess to compensate for losses.

134. Permissible charging rate of Edison batteries. In the case of the Edison battery the cells may be charged at any rate which does not cause the temperature of the solution in the cells near the center of the battery to rise above 115 deg. fahr.

135. Permissible charging rate of lead batteries. In the case of the lead battery, if the charging rate is not excessive, practically all the current is utilized for charging the plates. If the charging rate is increased, a point is reached where gassing begins, caused by bubbles of oxygen and hydrogen formed at the surface of the plates by decomposition of water in the electrolyte. This gassing increases with further increase in the charging rate. The excessive portion of the current which causes gassing produces no useful effect in charging the plates and is wasted. These high charging rates, which produce violent gassing, are not only wasteful of electric energy but tend to dislodge the active material from the plates, producing an excessive temperature rise, and materially shortening of the life of the plates. In general, any charging rate is permissible which does not produce excessive gassing or a cell temperature exceeding 110 deg. fahr.

136. Factors which affect gassing. The value of the charging current at which appreciable gassing begins in lead cells depend upon several factors such as the state of charge, temperature, specific gravity of electrolyte, type of plate, etc., but the principal factor is the state of charge of the battery. When the battery is fully charged any rate of charge however small will produce gassing, but the rate may be reduced to such a low value that, unless abnormally prolonged, the small amount of gassing which results is practically harmless. This safe rate is called the "finishing rate." During the early stage of the charge, the charging rate may be several times the finishing rate without producing violent gassing and the more completely the battery is discharged the higher may be the charging rate without causing excessive gassing.

137. Method of charging to prevent gassing. It has been found that if the charging rate throughout the charge of a lead battery is constantly adjusted to a value equal in amperes to the number of ampere hours out of the battery at any instant, violent gassing and excessive rise of temperature will be avoided.

138. Three common methods of charging. There are three common methods of charging batteries, as follows:

(1) The constant-current method.
(2) The constant-potential method.
(3) The modified constant-potential method.

139. The constant-current method. In the case of Edison cells this method involves the passing of a uni-directional current through the battery at the normal rate throughout the charge, without change. With lead cells the starting rate is maintained until the charge has progressed sufficiently to cause slight gassing or until the voltage has risen to a value of approximately 2.5 volts per cell; at this point the current is reduced to the finishing rate by manipulating the field rheostat of the generator or increasing the resistance in the charging rheostat. In order to maintain the proper rate it is necessary, at periodic intervals during the charge, to adjust either the generator field rheostat or the charging rheostat. The constant-current method is inherently a non-automatic method, since an adjustment of rheostat is necessary to maintain the constant rate.

140. Constant-current method for Edison batteries. The constant-current method has been recommended as best suited for charging Edison batteries, because it is more flexible and more easily handled by the operator.

141. The constant-potential method of charging is the method wherein the battery is connected, without a series rheostat, to a constant supply voltage, the particular value of which depends upon the battery to be charged. For a lead battery, the supply voltage should be approximately 2.3 to 2.4 volts per cell and for an Edison battery approximately 1.7 volts per cell. This method of charging, however, involves a considerable increase in the initial charging rate, which in many cases would necessitate excessive carrying capacity in charging apparatus and conductors. It requires considerable care in adjusting the bus voltage to meet the changes in temperature at different seasons and the charging rate may be subject to undesirable variations with slight fluctuations in bus voltage.

142. Constant-potential method for Edison batteries. This method may be used for charging Edison batteries under certain specific conditions, which should be determined by competent engineers. Only batteries in approximately the same state of discharge can be charged from a bus-layout which has been designed for this specific condition.

143. Advantages of constant-potential method. This method of charging has some advantages over the other methods and also some inherent disadvantages. It is the most efficient method, there being no resistor in the charging circuit and consequently no rheostatic losses. The control equipment is the simplest and the battery is charged automatically, the attention of an attendent being unnecessary.

144. Automatic charging. Any number of batteries up to the capacity of the supply circuit may be charged from the supply source. Automatic termination of the charge can be secured by the application of an ampere-hour meter and a magnetic contactor for each individual circuit. It is also possible, by the proper design of control equipment, to charge any number of batteries from a motor-generator set of proper capacity and secure full automatic operation of the entire system. A failure of the motor supply voltage at any time during the charging period will cause the batteries to be disconnected individually and upon the reestablishment of the supply voltage the motor-generator set will be automatically started and the charging of the various batteries automatically resumed. As the batteries become charged, each will be automatically disconnected from the circuit and the disconnection of the last battery will cause the motor generator set to be shut down.

145. Disadvantages of constant-potential method. The high initial charging current necessitates larger capacity of generating and control equipment. A small fluctuation in the supply voltage will affect the charging rate very materially, which makes it necessary to maintain the generator or line voltage at the specific required intensity throughout the charging period. If the supply voltage is only slightly in excess of that required, in charging a lead battery, counter-e.m.f. cells may be connected in series with the battery to reduce the voltage to the proper value. Counter-e.m.f. cells produce an opposing voltage of approximately 3 volts per cell. A resistor in series with the battery will not secure the same results as counter-e.m.f. cells.

146. The modified constant-potential method is the simplest of automatic charging systems, to which, fortunately, the characteristics of the lead storage battery are ideally adapted. Any method that involves manual adjustment introduces additional expense as well as a human element, which is often difficult to control; therefore the automatic charging of vehicle batteries has great advantages. This system consists of either a constant-voltage direct-current bus with a suitable fixed resistance in each charging circuit, or an individual charging machine for each circuit with a drooping characteristic equivalent to that obtainable with the fixed resistance. In either case an ampere-hour meter is arranged to disconnect the battery from the charging circuit upon completion of the charge.

147. Characteristics of modified constant-potential method. This method produces a comparatively high rate of charge at the beginning, which is gradually and automatically tapered as the charge progresses so as never to exceed a safe value. Therefore it is necessary to determine only the proper values for the constant bus-voltage and the fixed resistance to produce the results desired, namely: (1) To complete the charge within the time available; (2) in the case of lead batteries, to end the charge at the finishing rate or less; (3) to avoid a charging current in excess of the capacity of the charging source or the charging circuits.

148. Length of charging period. In fixing the time for completing the charge (Par. **147**), advantage should be taken of all the time available, as this will result in lower charging rates, smaller capacities for charging apparatus and circuits and lower cell temperatures. The shorter the time available, the higher will be the initial charging rate.

149. Application of modified constant-potential method to lead batteries. In the case of a lead battery, any combination of constant bus-voltage of not less than 2.4 volts per cell and a fixed resistance which will complete the charge at or below the finishing rate will produce safe charging rates throughout the entire charge.

150. Application of modified constant-potential method to Edison batteries. With Edison batteries the time required for a full charge by this method is the same as that required by the constant current method. The bus voltage should not be less than 1.8 volts per cell.

151. Voltage of charging bus. If the bus voltage is subject to sustained variations, it s necessary with lead cells that the conditions be such that the finishing rate of charge will not be exceeded when the bus voltage is at its maximum value, and yet that the charge be completed in the available time, which will depend on the average bus voltage. For any given number of cells, the higher the bus voltage the greater will be the loss in the resistance and therefore the lower will be the efficiency; also, the longer will be the time required to complete the charge of lead cells without exceeding the fin'shing rate at the end. On the other hand, if the bus voltage is but little in excess of that of the battery, the charging current will become more or less unstable and subject to considerable variation with variations of battery temperature or slight changes in the bus voltage.

152. Best value of bus voltage per cell. For lead cells, when the time available for charging is limited or efficiency is of considerable importance, a bus vo'tage equal to approximately 2.6 volts per cell will be found generally satisfactory and is recommended. The initial charging current will have several times the value of the finishing rate and will taper off as the charge progresses until it drops to, or below, the finishing rate at the end of the charge.

153. Effect of high-bus voltage and resistance. As the bus voltage and resistance are increased, the number of cells remaining the same, the charging current becomes more nearly constant throughout the charge, thus approaching the constant-current method.

154. Bus voltages lower than 2.6 volts per cell, modified constant-potential method may be used with corresponding reduction in the fixed resistance until the bus voltage is reduced to 2.3 to 2.4 volts per cell, with no resistance in circuit except that of the charging leads.

155. Best length of charging period. A charging period of approximately 12 hr. is required to charge a lead battery when the line voltage is 2.9 volts per cell. When the voltage exceeds this value the charging period will be increased and therefore the time available wil' probably be the limiting factor, if efficiency is not considered. The disadvantage of a very long charging period under high voltage conditions can be practically obviated by the use of control equipment for the purpose of inserting additional resistance in the charging circuit to reduce the current to the finishing rate. In the case of Edison cells the flatter the current-time curve produced by the higher supply voltage per cell the more desirable the charging characteristics will become, but the accompanying decrease in efficiency will probably be the limiting factor.

156. Modified constant-potential charging is the more desirable of the two non-variable voltage methods for either lead or Edison batteries. It embodies all the features of automatic charging in the constant-potential system and reduces the required capacity of the charging apparatus and the conductors. The charging resistors may be mounted either on the charging sections and wired to the face plates, in order to secure close current adjustment at the start of the charge, or they may be mounted apart from the switchboard. Taps should be provided on the resistors for close adjustment when mounted separately.

157. Equalizing charge. It is desirable occasionally to give a lead battery an equalizing charge by prolonging the ordinary charge until it is quite certain that all the cells and all the plates in each cell are fully charged. It has been customary to do this at various rates below the finishing rate, down to one-half, as a precaution against excessive temperatures. However, if it is not practical to provide means for reducing the rate, the finishing rate may be used for the equalizing charge, provided the charge is interrupted when necessary to prevent the temperature from exceeding safe limits.

158. Boosting charge. Any battery may be boosted at high charging rates during brief periods of idleness, thereby materially adding to the charge, provided the temperature of the solution in cells near the center of the battery does not exceed 110 to 115 deg. fahr. The following figures may be used under average conditions for Edison batteries, but values that will not cause excessive heating must be determined in each case by experience: Five minutes at five times normal rate; 15 minutes at four times normal rate; 30 minutes at three times normal rate; 60 minutes at two times normal rate.

Frothing at the filler opening is an indication that the boosting has been carried too far (if the solution is at the proper height) and the high rate should be discontinued at once.

Lead-battery boosting rates may be determined by the following equation:

$$\text{Rate} = \frac{\text{ampere-hours discharged from battery}}{1 + (\text{time in hours available for boosting})}$$

The charging rate in amperes multiplied by the time in hours and by 90 per cent. will give the total ampere-hour capacity which can be added in this manner to the normal or rated ampere-hour capacity, their sum being the total available ampere-hour capacity during operation.

159. Automatic termination of charge can be most satisfactorily obtained by the use of an ampere-hour meter in the battery circuit in conjunction with a shunt-trip circuit breaker or magnetic contactor in the charging circuit, connected in such relationship that when the ampere-hour meter needle is returned to the zero position at the completion of the charge, the circuit is completed through the operating coils of the circuit breaker or contactor in a manner such as to cause the device to trip, disconnecting the battery from the supply circuit. A shunt-trip circuit breaker or magnetic contactor has auxiliary contacts which open the tripping circuit when the breaker or contactor functions, thereby avoiding arcing at the ampere-hour meter contacts. The circuit breaker may be located either on the vehicle or on the charging panel and the magnetic contactor is usually located on the charging panel; in the last two cases a third conductor between the battery and the charging panel will be required to complete the tripping circuit. A three-point charging plug and receptacle are available in which the third point is designed for completing the connections of the trip circuit of the ampere-hour meter.

BATTERY CHARGING EQUIPMENT

160. The choice of charging equipment is determined by the conditions as to the source of energy, which may be classified as follows:

I. Direct-current constant-voltage source:
1. Use existing supply circuit if the voltage is suitable.
2. Reduce to suitable voltage by:
   (*a*) Counter-e.m.f. cells for charging lead cells
   (*b*) Balancer set
   (*c*) Motor generator

II. Alternating-current constant-voltage source:
1. Convert to direct current at suitable voltage by:
   (*a*) Motor-generator set.
   (*b*) Synchronous converter.
   (*c*) Mercury-arc rectifier.
   (*d*) Rectifier of Tungar type.

III. Special cases:
   (*a*) Charging for limited time from high-voltage direct-current source with ampere-hour meter control.
   (*b*) Single battery charged from motor-generator or other converting apparatus with drooping characteristics.
   (*c*) Several batteries charged in series from high-voltage bus.

161. Preference for direct-current source. When a direct-current source of energy is available it should be used if practicable. This will obviate the investment required for transforming apparatus and will often dispense with attendants who would be required if rotating machinery were installed.

162. Reduction of direct-current supply voltage. If it is necessary to reduce the voltage of the existing direct-current supply circuit, this may be accomplished in a number of ways, depending on conditions, as briefly described in Par. **163** to **168**.

163. Counter-electro-motive-force cells may be used for charging lead cells if the required voltage reduction is not more than 12 per cent. of the bus voltage. These cells are similar to storage-battery cells (in glass jars or lead-lined tanks), except that the plates have no capacity for discharge. The function of the counter-e.m.f. cell is to absorb a portion of the line voltage and thereby adjust the voltage impressed on the battery terminals. In counter-e.m.f. cells the voltage drop is nearly constant, whereas in a resistor the voltage drop is directly proportional to the current. By the use of the counter-e.m.f. cells, the characteristics of the modified constant-potential method may be secured with a bus voltage somewhat higher than normal and without manual adjustment during charge, whereas if an ordinary rheostat is used, adjustments must be made from time to time to secure equivalent results.

164. The voltage absorbed by a counter-e.m.f. cell varies from 2.3 volts at low current to 3 volts at high current. For continuous service, the current flowing through a counter-e.m.f. cell should be limited to the 3-hr. rate of the corresponding storage cell, but higher rates than this, up to the 1-hr. rate, may be used for short periods or for the initial stage of a tapering charge lasting 8 hr. or less.

165. Auxiliary bus derived from counter-e.m.f. cells. These cells may also be used to provide a second constant-voltage bus, at lower voltage than the main bus, for charging batteries of fewer cells.

166. Use of balancer sets. A balancer set may be used if the bus voltage is approximately twice the value required for charging. This equipment is comprised of two identical direct-current machines mounted on a common shaft and electrically connected in series across the main bus, thus providing a three-wire system, each side of which comprises a suitable source of supply for the modified constant-potential system. If the line voltage is slightly more than twice the value suitable for lead cells, the balancer set can be used in conjunction with counter-e.m.f. cells.

167. Capacity of balancer sets. Where a charging source of double the desired voltage is already available and of ample capacity for all future requirements, a balancer set of sufficient capacity for the maximum demand of the largest single battery will be satisfactory for any number of additional batteries comprised of the same number of cells. The balancer set is required to handle only the unbalanced load, which is the difference in load on the two sides of the three-wire circuit; this difference need never be greater than the maximum demand of the largest battery.

168. A direct-current motor-generator set should be installed if the direct-current bus voltage is either too low or too high and not suitable for a balancer set. This equipment consists merely of a direct-current generator direct-connected to a direct-current motor, with a common bed-plate; belted sets, while practicable, require more room and are more expensive to maintain.

169. Alternating-current motor-generator set for charging Edison batteries. This type of installation for charging an Edison battery from an alternating-current source requires a motor-generator set and a single-circuit automatic control panel. A differential-wound generator has characteristics which make it suitable for charging an Edison battery automatically, inasmuch as it is practically a constant-current machine and maintains a constant charging rate throughout the charge. An individual generator is required for each battery. If automatic termination of the charge is desired, a magnetic contactor and ampere-hour meter must be provided in the control circuit of each battery. This equipment will require no attention after the charge is started. The control equipment can be so designed that it will disconnect the battery from the generator upon completion of the charge and also disconnect the motor from the main supply circuit.

170. Alternating-current motor-generator set for charging lead batteries. Similar equipment (Par. **169**) can be used for charging lead cells automatically, by providing the proper control equipment and an ampere-hour meter having an additional or intermediate contact. This intermediate contact is arranged to connect an additional resistance into the shunt-field circuit for the purpose of reducing the charging current to the finishing rate when the charge has reached a point within approximately 20 per cent. of completion. In the case of lead batteries the expense of supplying an individual generator and control panel for each battery is not warranted, because they are so well adapted to the modified constant-potential charging method.

171. Charging equipment for commercial trucks. An installation serving two or more commercial trucks will require a motor-generator set, with a generator control section and one charging section per truck. An installation serving two industrial trucks may be equipped with two individual motor-generator sets with automatic control panels, with capacity for one battery each, or with one motor-generator set for both batteries, with generator control panel embodying two charging circuits. An installation serving three trucks or more will require a motor-generator set, with automatic control panel providing one charging section per truck.

172. Charging equipment for passenger cars. For private passenger cars where there is one car in a private garage a mercury rectifier, a Tungar type of rectifier, or a motor-generator set and automatic control panel will be satisfactory. Where there are two or more passenger cars in a public garage a motor-generator set and generator control panel with one charging section for each battery to be charged (at the same time), and also one or two mercury-arc rectifiers for light load conditions, should be satisfactory.

173. Quick charge from high-voltage bus. It will be noted that as the bus voltage increases the minimum time for completing the charge of a lead battery through a fixed resistance, without exceeding the finishing rate, becomes longer. If it is desired to hasten this charge, it can be accomplished without abandoning the automatic feature by using an ampere-hour meter with an extra auxiliary contact and two fixed resistances in parallel in the charging circuit. The charge is started at a high rate through the two resistances and when the charge is partly completed the extra auxiliary contacts on the ampere-hour meter open an automatic switch and cut out one of the resistances, after which the charge is continued at a reduced rate through the remaining resistance until completion, when the charge is finally terminated by the other auxiliary contacts on the meter.

174. The extra auxiliary contact on the ampere-hour meter is located at a distance from the zero point equal to 20 per cent. of the total dial range of the meter and is designed so that when the hand of the meter passes it on discharge it closes the contact, and it remains closed until the meter hand opens it when passing in the opposite direction during charge. When connections are made for the charge, this contact being closed, the automatic reclosing breaker which controls the second resistance also closes and the charge continues at the high rate until at the proper time the meter opens the contact and thereupon the breaker opens, thus reducing the rate. This device, while somewhat more complicated than the shunt-trip breaker, offers more complete protection for the reason that if any trouble develops in the extra contact or in its circuit the automatic breaker immediately opens, or fails to close, and thereby prevents excessive charging rates during latter part of the charge.

175. Adaptability of meter to different sizes of battery. It is necessary to use one size of ampere-hour meters for as many as four different sizes of battery, so that it is commercially impracticable to have the extra auxiliary contact on the meter adjustable for each of these battery sizes. The location chosen, at 20 per cent. of the dial range, gives full production under the most severe conditions, namely, when the largest battery is being charged in the shortest available time. When a smaller battery is used with the same size of meter or when a longer time is available for charging, the current will be reduced earlier and therefore a longer time will be required to complete the charge than would be necessary if the location of the contact could be adjusted to suit those lower charging rates.

176. Auxiliary resistance equal to normal resistance. Designating the fixed resistance to reduce the final rate of charge to the finishing rate as the "normal" fixed resistance and the time required to complete the charge through this resistance alone as the "normal" charging time, then if the second or auxiliary resistance is equal to the normal, the charging rate during the first stage of the charge will be approximately doubled. The second resistance should have the same carrying capacity as the first. The charge will then be completed at the finishing rate through the normal resistance and the total time required will be from 64 to 73 per cent. of normal.

177. Auxiliary resistance equal to one-half of the normal resistance. If the second resistance is one-half normal, the charging rate during the first stage will be three times normal, and the second resistance should have a carrying capacity double that of the first. The total time required will be from 50 to 63 per cent. of normal.

178. Further reduction in the second charging resistance, to less than half of the normal resistance, cannot be made without encountering excessive charging rates during the initial period of the charge.

179. Main and control conductors. This arrangement requires two control conductors from the ampere-hour meter contacts to the charging panel, in addition to the main conductors, and the charging plug and receptacle must have two auxiliary contacts in addition to the two main contacts. If the circuit breakers and a part of the charging resistance can be mounted on the vehicle, only the two main conductors and a two-point charging plug and receptacle will be required.

180. Single battery charged from motor-generator or other converting apparatus with drooping characteristic. If a single battery is to be charged from a motor-generator set or other transforming apparatus, results the same as those obtained from the modified constant-potential method may be secured, without the use of the fixed resistance, by designing the generator or converter apparatus with the proper drooping characteristic. To accomplish this the value of the fixed resistance and the bus voltage for the modified constant-potential method must first be determined and then the charging apparatus must be designed for a no-load voltage equivalent to the constant bus voltage, and the drop in voltage at full load should be the same as the drop in the fixed resistance with the same current flowing. Characteristic curves of such a type of generator are given in Fig. 18.

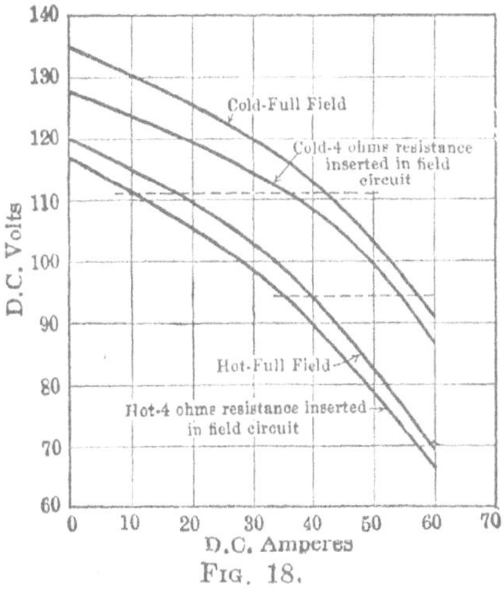

Fig. 18.

181. Charging batteries in series. In cases where there are several batteries to be charged from a high-voltage direct-current bus, it is sometimes the practice to connect the batteries in series across the bus through a suitable fixed resistance. This method will prove satisfactory if all the batteries are discharged to approximately the same state.

182. Precautions in charging batteries in series. In the case of lead batteries, if they are likely to be unequally discharged, the charging rate should be limited to the finishing rate throughout the charge and means should be provided for cutting out any battery that becomes fully charged in advance of the others, connecting in its place a fixed resistance which will produce a voltage drop equivalent to the voltage consumed by the battery cut out.

183. Special types of motor-generator sets for converting alternating-current energy to direct current of suitable voltage are available in various forms and combinations.

184. Individual taper-charging motor-generator sets are in general divided into two classes, low-voltage sets for charging the batteries of industrial trucks and tractors and high-voltage sets, for charging the batteries of trucks and passenger cars. Each set is designed for charging one battery only.

185. Taper-charging generators are specially wound to provide a tapering charge automatically. For a lead battery, and also in the case of an Edison battery, the tapered charging rate is adjusted so that the average current value is equivalent to the normal constant-current rate for the particular battery.

186. Constant-current charging motor-generator sets. Motor-generator sets are also available having their generators differentially wound, in order to maintain automatically an approximate current rate suitable for charging Edison batteries. These sets are capable of automatically charging either one or two Edison batteries having the same electrical characteristics, without the services of an attendant to adjust the field rheostat or the battery-charging rheostat to obtain the proper charging rate. As the charge progresses and the battery-e.m.f. rises, the generator-e.m.f. also rises due to the differential action of the fields, automatically maintaining a current rate which is approximately constant.

187. Multiple-battery charging motor-generator sets are designed for charging two or more vehicle batteries in multiple. They are furnished in various voltages and capacities for any of the standard sizes of vehicle batteries. A complete range of sizes for charging batteries of the Edison and the lead types is available, having either shunt-wound or compound-wound generators. From 2.5 to 30 kw. capacity the generators are obtainable in three different ranges of voltage rating, namely 55 to 35 volts, 80 to 51 volts and 100 to 76 volts. The generators can also be obtained in the standard 125-volt classification.

188. Control panels for individual taper-charging motor-generator sets are obtainable for either manual or automatic control, arranged for wall mounting. The panel, which is of slate, is equipped in the non-automatic type with a triple-pole fused knife-switch with special starting clip, fuses in generator and motor circuits, starting resistance, generator field-rheostat and necessary ammeter and voltmeter.

189. The automatic type of panel is used in connection with a contact-making ampere-hour meter. This panel is equipped with a triple-pole circuit-breaker with shunt-trip attachment, which will automatically open both motor and generator circuits when the battery has been charged to an amount predetermined by the setting of the ampere-hour meter (see Par. **159**). This panel is also equipped with ammeter, voltmeter, fuses, generator field-rheostat and special spring-opening switch for starting the set from the battery.

190. Control panels for multiple-charging motor-generator sets are available in various forms and combinations, but the different types are too numerous to permit of detailed description. The general features are covered in the following paragraphs.

191. Sectional type of panel. The panels of sectional type can be assembled in a large variety of possible combinations, thus making it possible to design a very flexible installation. The number of charging circuits may be increased at any time after the switchboard has been installed by the addition of suitable sections and rheostats. However, the combined capacity of the charging circuits in use at one time must not exceed the capacity of the main section. Each panel is comprised of a generator section, an instrument section, and one or more charging sections, together with the necessary charging rheostats. The equipment provides for protection against reverse current and reverse polarity; when an ampere-hour meter is used it also provides for disconnecting the battery from the generator upon completion of the charge. A switchboard of this type centralizes the control of a large number of batteries and is convenient to operate.

192. Mounting of battery-charging rheostats. In general the battery-charging rheostat is comprised of a resistor of cast-iron grids supported on the rear of each panel section between steel end-frames, with contact buttons and moving arms on the front of the panel section. Bolts through the panel hold the end frames of each charging rheostat to the panel, so that the rheostat forms an integral part of the section equipment.

193. Panel framework and assembly. The sections of each panel are assembled one above another and securely bolted to a vertical angle frame, or other type of frame, of suitable height. It is not advisable to use panels exceeding 84 in. in height because some of the switching apparatus would be inconveniently high for the operator. When more sections are required than can be mounted on a frame 84 in. high, they may be arranged in two or more panels of uniform height. Where the necessary number of sections is not sufficient to make all panels of the same height, blank sections may be added to secure uniformity.

194. Typical panel connections. A typical diagram of connections of a sectional panel for the control of a motor-generator set is shown in Fig. 19.

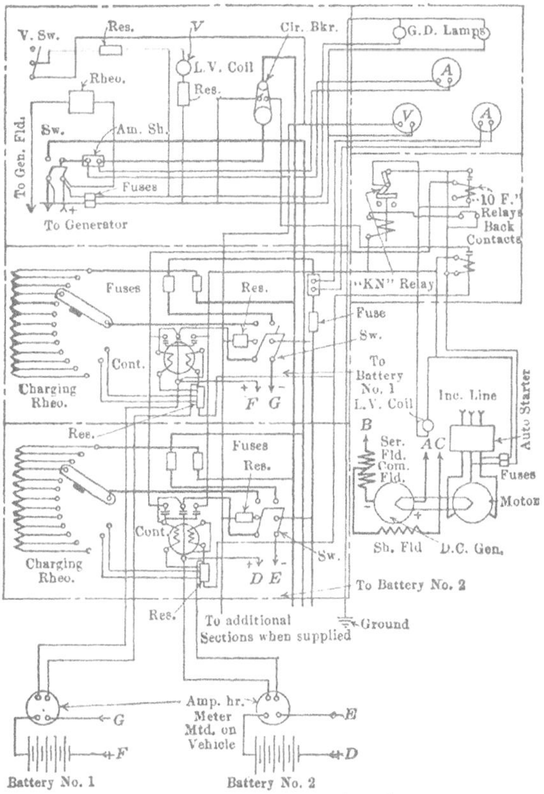

FIG. 19.—Diagram of connections of a sectional type charging panel for control of a motor generator set.

195. Panel connections for automatic charging. A sectional type of panel is also available, with suitable relay equipment, for automatically terminating the charge and shutting down the motor which drives the battery-charging set. The zero contacts of the ampere-hour meters, of which there is one in each charging circuit, short-circuit the holding-in coils of the battery line contactors. This action causes the battery line contactor to drop out and opens the corresponding circuit. Each battery is disconnected successfully, in similar manner, as its charge is completed. The tripping of the last charging contactor opens a relay circuit connected to the low-voltage coil of the auto starter, which causes the starter to disconnect the motor-generator set from the line. A typical diagram of connections of this type of panel is given in Fig. 20.

196. Panel connections for automatically charging a single battery. For the control of a generator where only one battery is to be charged at a time, and where automatic termination of the charge is desired, together

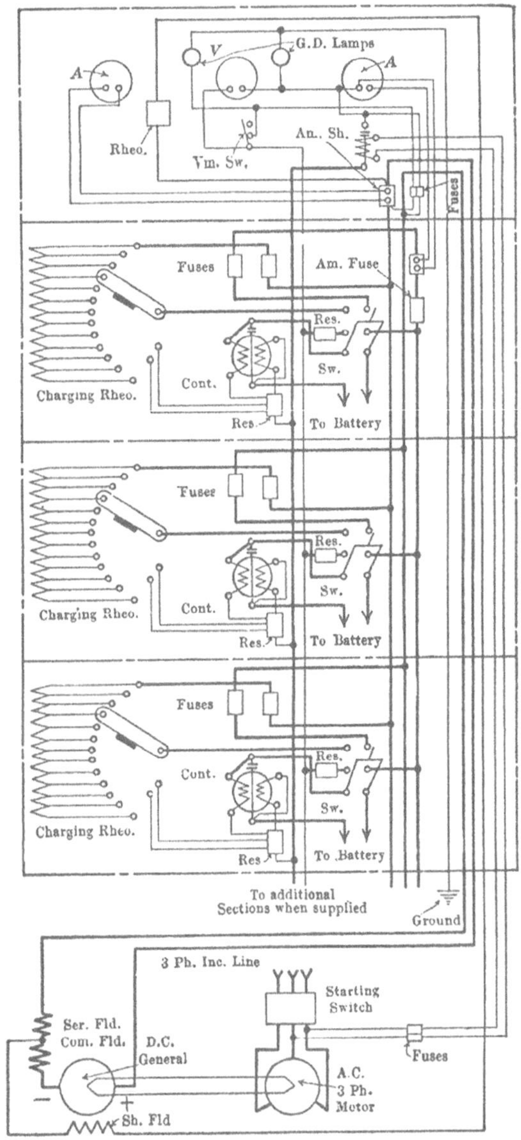

FIG. 20.—Diagram of connections of a sectional type charging panel for control of a motor generating set for full automatic charging.

with low-voltage and reverse-current protection and automatic shut-down of motor-generator set on completion of charge, the typical panel connections are shown in Figs. 21 and 22.

197. Panels for direct-current supply, of suitable voltage, are similar to the sectional types previously described for the control of multiple-charging motor-generator sets, the principal difference being the substitution of a line and meter section for the generator section. A typical diagram of connections is shown in Fig. 23.

198. Panels for automatic charging from direct-current supply. Where only one battery is to be charged at a time and where full automatic

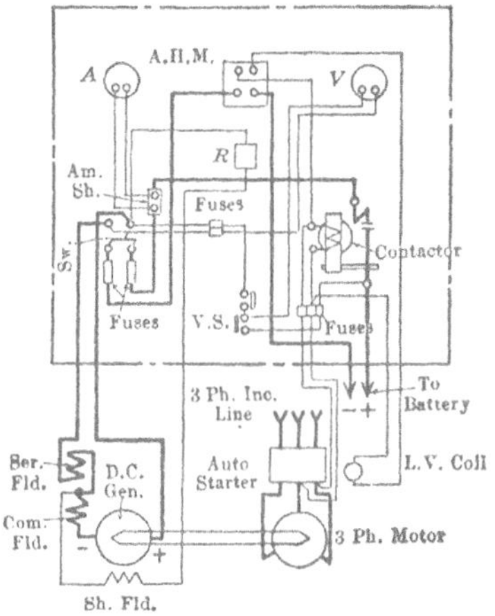

Fig. 21.—Charging an Edison battery from a constant current or constant potential generator.

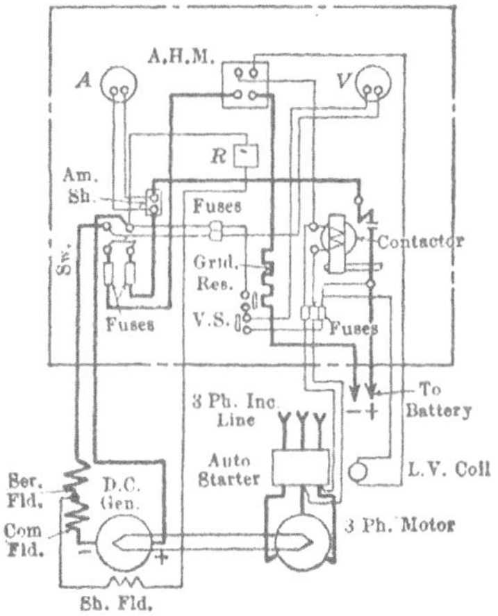

Fig. 22.—Charging a lead battery from a constant potential generator.

199. Table of Data on Battery-charging Rheostats for Circuits up to 115 Volts Maximum

(General Electric Company)

Charging amperes		Type of cell	No. of cells	No. of steps	Total ohms
Start	Finish				
15	5	Lead	12–18	15	17
20	6	Lead	12–18	15	17
20	5	Lead	20–28	15	13
25	8	Lead	20–28	15	9
30	10	Lead	30–36	15	4
40	10	Lead	30–36	15	4
30	10	Lead	37–40	15	2.6
40	10	Lead	37–40	15	2.6
30	10	Lead	41–44	15	1.5
40	10	Lead	41–44	15	1.5
50	12	Lead	41–44	15	1.28
60	15	Lead	41–44	14	1.0
30	30	Edison A-4	20–40	15	2.8
30	30	Edison A-4	44–60	15	1.7
45	45	Edison A-6	20–40	15	2.0
45	45	Edison A-6	44–60	15	1.2
60	60	Edison A-8	20–32	14	1.4
60	60	Edison A-8	36–44	14	1.0
60	60	Edison A-8	48–60	14	0.72
75	75	Edison A-10	20–32	10	1.17
75	75	Edison A-10	36–44	10	0.81
75	75	Edison A-10	48–60	10	0.60
90	90	Edison A-12	20–32	10	0.96
90	90	Edison A-12	36–44	10	0.68
90	90	Edison A-12	48–60	10	0.48

charging is desired, the control panel will be similar to that described for the control of the generator in Par. **196.** Typical connection diagrams are shown in Figs. 24 and 25.

200. Capacity of battery-charging rheostat. Rheostats for battery charging are specially designed to meet the various conditions fixed by the

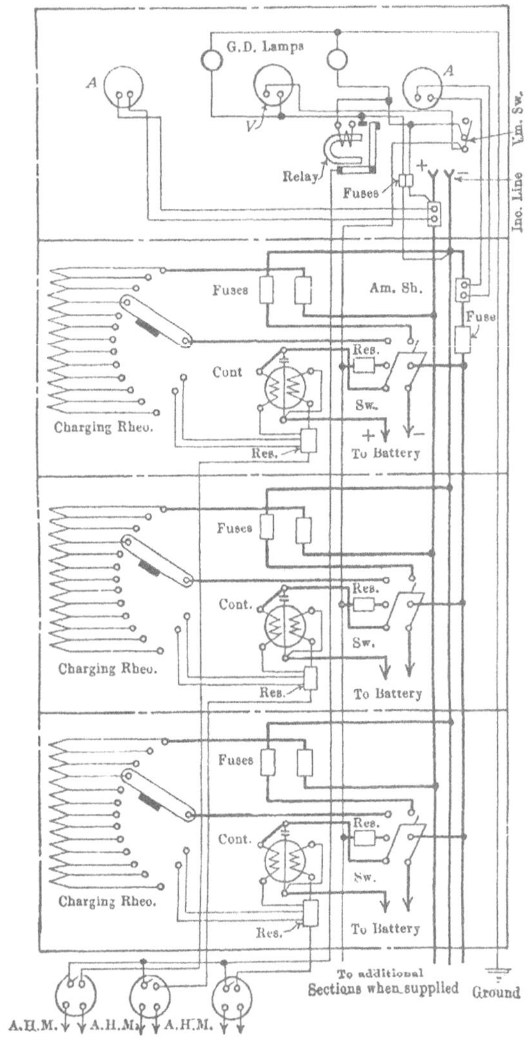

FIG. 23.—Diagram of connections of a sectional type charging panel for control of an incoming D. C. line.

number and type of cells in the battery to be charged and the voltage of the direct-current supply circuit. Only by using the proper rheostat for a given battery is it possible to start the charge at the correct rate and reduce the current in suitable steps to the final required rate (Pars. **199** and **201**).

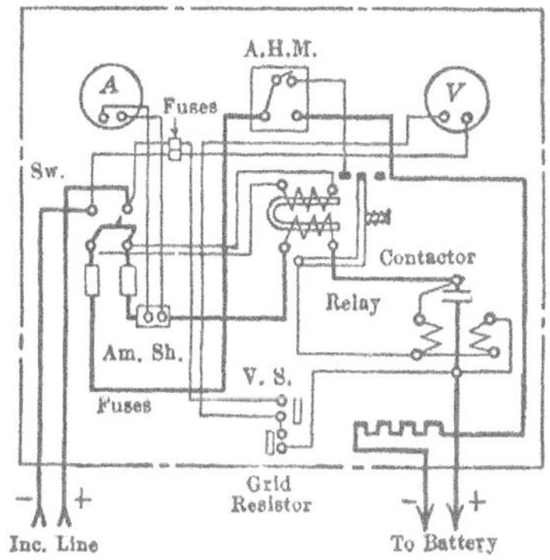

FIG. 24.—Charging a lead battery.

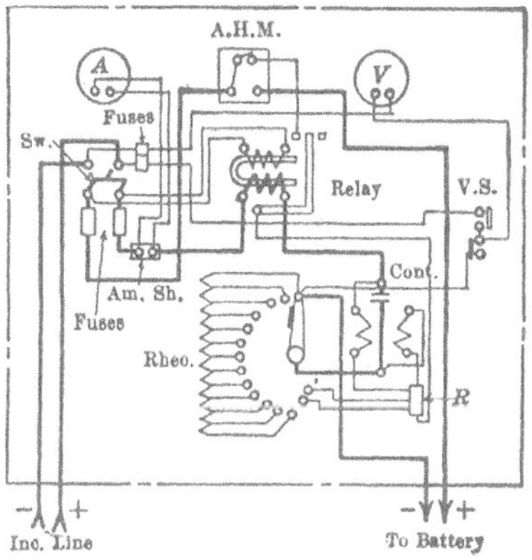

FIG. 25.—Charging an Edison battery.

201. Table of Data on Special Battery-charging Rheostats for Edison Cells, at Normal or Double Normal Rate

	Charging amperes		Type of cell	Number of cells	Number of steps	Total ohms
	Double rate	Normal rate				
Private garage	60	30	Edison A–4	60	14	0.56
	90	45	Edison A–6	60	14	0.34
	120	60	Edison A–8	60	10	0.26
	150	75	Edison A–10	60	14	0.235
	180	90	Edison A–12	60	14	0.23
Public garage	180	30	Ed. A–4–6–8–10–12	60	14	0.565
	120	30	Ed. A–4–6–8	60	10	0.63
	180	90	Ed. A–10–12	60	14	0.23

202. The mercury-arc rectifier and the Tungar rectifier are limited in capacity and can therefore be used only where the charging current is within this limit. Mercury-arc rectifiers can be obtained in capacities up to 50 amp., for any desired voltage. This type of rectifier has the disadvantage that the charging current must be maintained above a certain minimum value and it is not ordinarily self starting. The mercury-arc rectifier has a more or less drooping characteristic, that is, the terminal voltage drops with increased output.

203. Mercury-arc rectifier installation. The general appearance of the rectifier equipment including the reactance, rectifier tube, panel, switches, and instruments is shown in Fig. 26.

204. Non-automatic type of rectifier. One type of rectifier is not automatic in operation, that is, it will not start or stop itself and if the line voltage fails temporarily, it will not restart. To start this type the bulb must be tilted by hand by means of a tilting handle.

205. Automatic rectifier. Another type has the same features as the non-automatic and in addition a tilting magnet for automatic starting and a relay circuit-breaker. In case of line-voltage failure, which will stop the rectifier, the cutout closes the tilting transformer circuit so that the rectifier is ready to restart itself when the power supply is restored. In case of excessive direct current, the cutout coil opens the circuit-breaker contacts.

206. Rectifier applications. For charging a single battery the characteristics of a rectifier are not objectionable. The rectifier is applicable where the current value does not exceed the rated capacity. The commercial ratings of mercury-arc rectifiers for vehicle-battery charging are shown in table in Par. **208**.

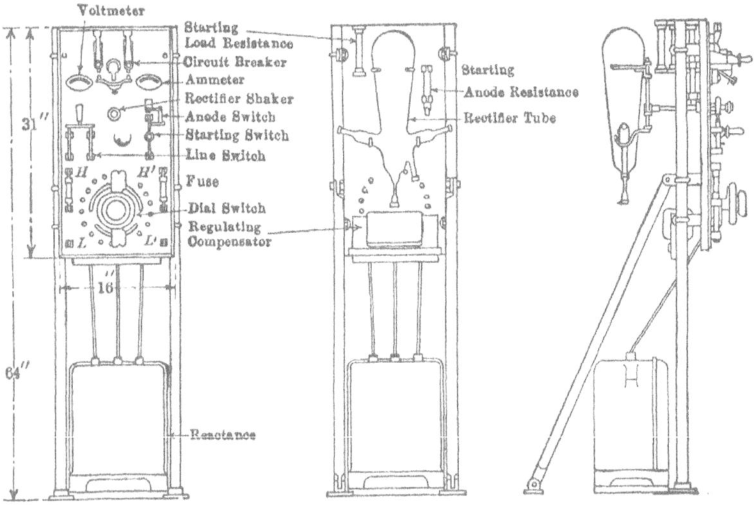

FIG. 26.—Mercury arc rectifier—front, rear and side elevations. (General Electric Company.)

207. Table of Commercial Ratings of Mercury-arc Rectifiers for Vehicle-battery Charging

Normal Frequency—60 cycles

Capacity, amp. (d.c.)	Make	Volts, range (d.c.)	Supply voltage (a.c.)	No. of cells		Approx. shipping weight (lb.)
				Lead	Edison	
30	G.E.	10– 75	110	5–30	7–40	590
	G.E.	20– 75	220	10–30	14–40	590
	West.	28– 85	110	14–32	....	485
	West.	28– 85	220	14–32	....	485
	G.E.	30– 83	110	12–32	16–44	540
	G.E.	30–118	220	14–46	16–62	540
	G.E.	10–100	110	5–38	7–54	590
	G.E.	20–120	220	10–46	14–65	590
	West.	40–120	110	20–44	....	575
	West.	40–120	220	20–44	....	575
	G.E.	75–175	220	32–68	54–95	590
50	G.E.	10– 75	110	5–30	7–40	660
	G.E.	20– 75	220	10–30	14–40	660
	G.E.	10–100	110	5–38	7–54	660
	G.E.	20–120	220	10–46	14–65	660
	G.E.	75–175	220	32–68	54–95	660

BIBLIOGRAPHY

208. Reference literature.

EAMES, HAYDEN.—"Methodical Machine Operation." Studebaker Co., 1908.

PENDER, HAROLD.—"Notes on the Cost of Motor Trucking." The Central Station; New York, 1912.

REID, H. C.—"Electricity as a Substitute for Horses in Local Removals." See American Warehousemen's Convention Report; Chicago, 1911.

KNOWLTON, H. S.—"Side Lights on the Electric Vehicle Situation." The Electric Vehicle and Central Station; Boston, 1909.

KENNEDY, W. P.—"The Motor Truck for Team Owners." Team Owners' Review, 1908. "Systematic Motor Vehicle Operation." Motor, 1909. "Economy of Motor Truck Transportation." Studebaker Co., 1909. "The Care of Electric Vehicles." Studebaker Co., 1909.

SMITH & CUSHING.—"The Electric Vehicle Handbook." Published by Smith & Cushing in New York, 1920.

THOMPSON, S. G.—"Scientific Elements of Automobile Transportation." The Electric Vehicle and Central Station; Boston, 1909. "Electric Vehicle Reliability." The Central Station; New York, 1911. "Electric Vehicle Application." The Central Station; New York, 1912. "The Electric Vehicle." The Central Station; New York, 1913.

WAGONER, P. D.—"European Development of the Electric Vehicle Industry."†

CARLE, F. NELSON.—"Special Applications of Electric Trucks."†

MANWARING, W. A.—"Power Wagon Operation in Central Station Service."†

DODGE, H. P.—"Constant Potential System of Charging from Motor Generators."†

LINCOLN, J. C.—"Charging Apparatus for the Private Garage."†

GROVE, ROBERT B.—"Electric Vehicle Performance."†

SCHOEPF, T. H.—"Effects from the Utilization of the Kinetic Energy of an Electric Vehicle."† "The Performance Characteristics of Electric Motors and Their Influence on the Operation of Electric Trucks." The Central Station, New York, Mar., 1914.

PENDER, H. and THOMPSON, H. F.—"Notes on the Cost of Motor Trucking," and "Observations on Horse and Motor Trucking," issued by Elect. Research Lab. of Mass. Inst. of Tech., 1912 and 1913.

LA SCHUM, E. E.—"The Electric Truck in Modern Transportation."†

NIMS, A. A.—"Calculations of Electric Motor Characteristics and Prediction of Vehicle Performance."†

BALDWIN, H. F.—"The Electric Automobile Motor."†

CHURCHWARD, A.—"The Standardization of the Electric Vehicle," Central Station, New York, 1912.

Handbook of the Soc. of Automotive Engineers, N. Y., 1920.

---

†*Proceedings* of Electric Vehicle Section, National Electric Light Association, New York, N. Y.

ELECTRIC VEHICLE AND ALKALINE STORAGE BATTERIES

by Walter E. Winship, PhD, 1922

ELECTRIC VEHICLE BATTERIES

120. Adaptable types. Pasted plates are universally used in vehicle batteries at the present time, although there is no reason why a battery with Planté positives and pasted negatives should not be successful on account of its greater ability to withstand "boosting" charges. By a boosting charge is meant a partial charge at a high current rate, such a charge taking place at the noon hour or other rest period after the battery is partly discharged. The Planté positive, even with a pasted negative, may be charged at enormously high rates without the injurious heating which develops in the Edison battery.

If in a particular service the ability of a battery to withstand boosting at high rates is of importance, this combination should receive consideration.

The prime reason for the use of pasted plates is their relative lightness, since a reduction in weight of battery results in a reduced current consumption for the vehicle and a reduction in the expense of tire upkeep. For these reasons, battery-plate life is often deliberately sacrificed by using thin plates and a small separation between plates. Small separation makes it necessary to employ strong acid and this reduces the life of the separators as well as the life of the plates. There is a strong tendency toward the use of thicker separators and a reduced acid strength

121. Plate dimensions. The size of plate in general use for vehicle purposes is $5\frac{3}{4}$ in. $\times$ $8\frac{5}{8}$ in. Standard positive plates vary from 0.22 in. to 0.13 in. in thickness, negative plates from 0.19 in. to 0.11 in. Several other sizes of plates are occasionally used, but they are gradually being abandoned.

FIG. 30.—Vehicle plate grid "shelf type."

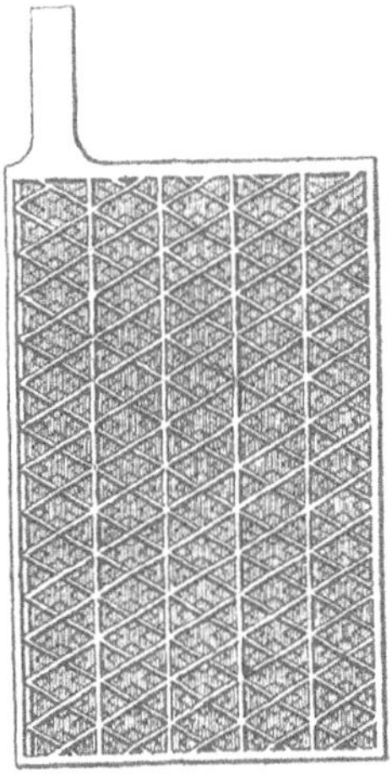

FIG. 31.—Vehicle plate grid "diamond type."

Negatives are usually made thinner than the positives with which they are used by several hundredths of an inch, so that the plates of each polarity shall have approximately equal life. Each battery manufacturer makes several thicknesses of plates, the proper thickness to use under any condition being dependent on the ampere-hour capacity required and the space and weight limitations. In a definite size of jar, a greater capacity can be obtained with a larger number of thin plates at high discharge rates. The cost of the thin-plate battery in the same size jar is somewhat greater, and the

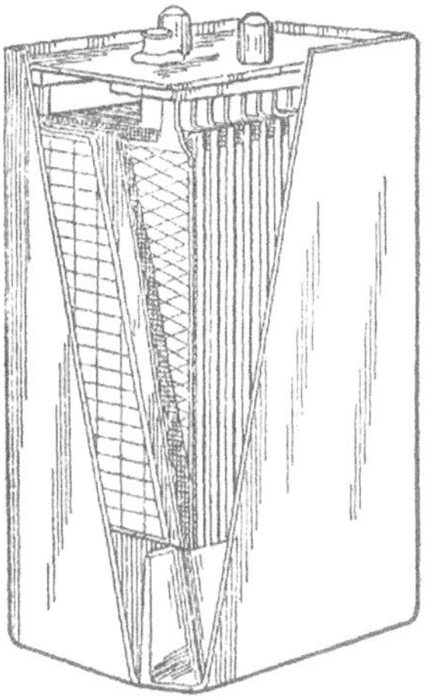

FIG. 32.—Vehicle cell construction.

ampere-hour life is slightly less. The ratios as obtained from bench tests, under exaggerated charging conditions, are shown in Par. **125**.

122. Choice of plate thickness. In the examples cited in Par. **125**, the prices and, also, the "probable life" are to be considered as comparative. The ratios hold for several makes of battery. Where maximum mileage radius of a vehicle is necessary, the thinner plates should be chosen. The best all-round results will be obtained, in the judgment of many, from using the intermediate plate thicknesses.

123. Grid types. The only two grid designs in general use with the ordinary plates are those illustrated in Figs. 30 and 31.

124. Vehicle battery construction is illustrated in Fig. 32. Jars are universally of hard rubber. Celluloid is not suitable for vehicle batteries on account of its inflammability. It also oxidizes in contact with positive plates and becomes worthless. The bottoms of jars contain supporting ribs or bridges, upon which the plates rest; these bridges provide a space beneath the plates to take the sediment which is gradually thrown off from the plates in service.

125. Relative Attributes of Plates of Different Thicknesses

Size of plates 5 in. × $8\frac{5}{8}$ in. Size of jars $6\frac{1}{4}$ in. wide × $4\frac{1}{4}$ in. long.

Type of plate	Weight per cell (lb.)	Capacity at 6-hr. disch. rate (amp.-hr.)	Approximate price to user (dollars)†	Average life (amp.-hr.)
Thick (heavy duty)..	36.8	150	11.00	75,150
Medium (standard)..	34.0	170	11.70	75,200
Thin (high capacity).	34.5	190	12.10	75,400
Extra thin (extra high capacity).....	35.8	200	12.75	67,200
Iron clad*...........	34.5	170	16.75	170,000

Size of plates $5\frac{5}{8}$ in. × $8\frac{5}{8}$ in. Size of jars $6\frac{1}{4}$ in. wide × $8\frac{1}{16}$ in. long.

Type of plate	Weight per cell (lb.)	Capacity at 6-hr. rate (amp.-hr.)	Approximate price to user (dollars)†	Average life (amp.-hr.)
Thick (heavy duty)..	70.0	305	19.50	155,000
Medium (standard)..	65.0	341	20.75	150,000
Thin (high capacity).	65.9	384	21.20	153,000
Extra thin (extra high capacity).....	64.0	420	22.30	141,000
Iron clad*...........	65.0	340	30.45	340,000

126. Separators. The plates are insulated from each other by thin sheets of corrugated wood veneer, grooved on one side, the wood used being either naturally free from such organic substances as would injure the plates, or else treated to remove such impurities. These veneers are usually kept away from the positive plate by inserting a thin sheet of perforated hard

* See Par. **75**. †The excise tax is not included.

rubber. The smooth side of the veneer is placed against the negative plate. The life of these wood separators is unfavorably influenced by very strong acid also by an excessive amount of overcharge. If batteries are properly operated, the separators will last as long as will the plates of a thin-plate battery, that is, 12 to 15 months in commercial service and 18 to 24 months in pleasure-vehicle service.

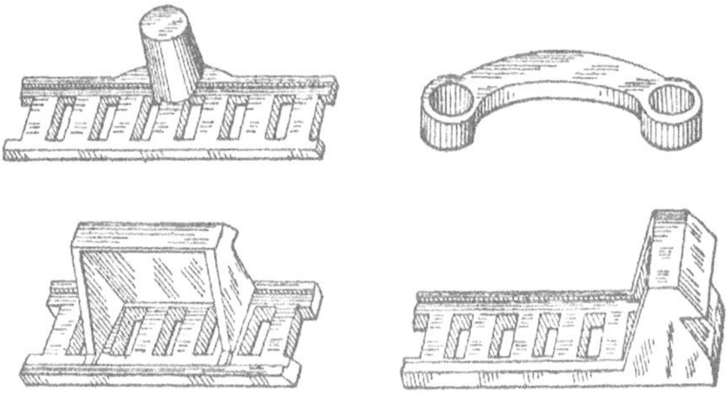

FIG. 33.—Vehicle type straps and connector.

Thick plates will generally have a longer life than the wood separators, so that, in order to obtain the best possible life of the plates, the separators will require renewal. If the separators and the plates are to give approximately equal lives, sufficient space should be provided in the jars to take all the sediment which will be thrown off in the life of the plates, *i.e.*, high-bridge jars should be used. Otherwise the space should be ample to take the sediment which will be deposited during the interval between separator renewals.

127. The straps to which the plates are burned are of several types, as shown in Fig. 33. The pillar-post strap, used with connector links, is generally considered to be the best design. Several kinds of bolted connections are on the market, but most of them are unsatisfactory. A burned connection is preferable.

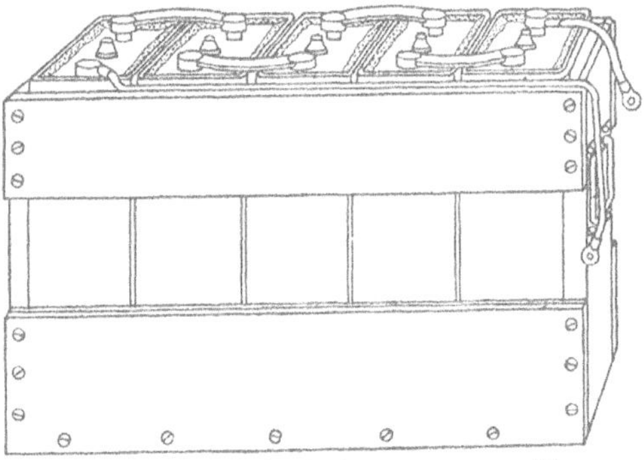

FIG. 34.—Vehicle battery assembly.

128. Covers. The cells are preferably provided with covers of hard rubber, sealed with an acid-resisting compound, to prevent slopping of the electrolyte, although frequently a tight-fitting cover is relied upon to prevent this. If the covers are sealed, **vent caps** of either hard or soft rubber are used.

129. Degrees of separation of plates. As it is desirable to reduce weight and space required, as far as possible, with vehicle batteries, separation between plates and the volume of electrolyte are limited; on this account the range of acid strength between the states of full-charge and full discharge is great. In extreme cases, acid of 1.300 sp. gr. at full charge is used, and only sufficient volume employed to give a final acid density of 1.100. A better life of plates and separators will result, however, if a wide enough separation between plates is used to permit a full-charge density of 1.260 to 1.280 sp. gr.

130. Crates. The method of assembling vehicle cells into crates is shown in Fig. 34. The crates are of hard wood, painted thoroughly with asphaltum acid-resisting paint or else soaked in hot paraffin oil, then subsequently dipped into melted paraffin. The practice of fastening crate terminals onto the crates is to be thoroughly condemned, as when the crates become acid-

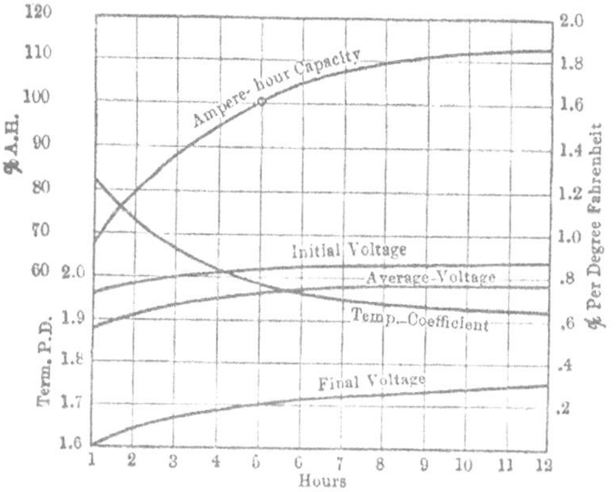

FIG. 35.—Capacity voltage and temperature characteristics.

soaked, there is current leakage between them, and the positive terminal corrodes seriously. Crate ends are charred from this cause, and the current leakage is of a magnitude not to be neglected.

131. Performance of thin-plate batteries. The variation of capacity of a thin-plate vehicle-type battery in wide commercial use, is shown in

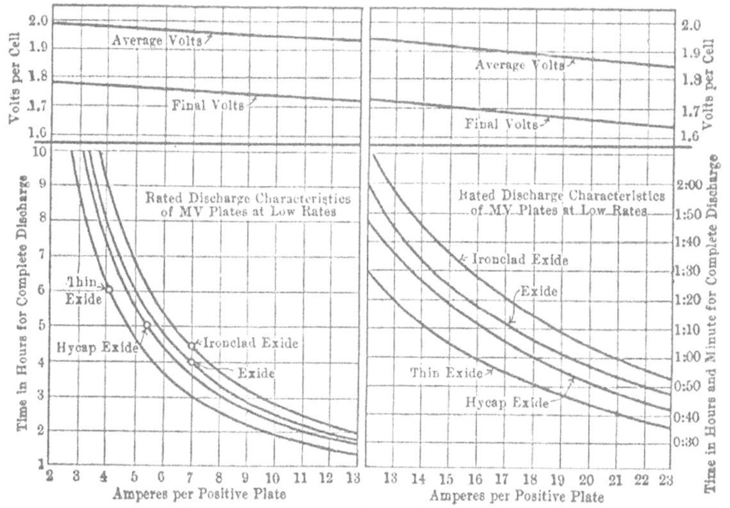

FIG. 36.—Exide plate characteristics.

Fig. 35, and the average and final voltages are also given. The Electric Storage Battery Co. has devised an ingenious chart showing the characteristics of their "Exide" batteries which is reproduced in Fig. 36.

ALKALINE STORAGE BATTERIES

213. Classification. There are but two practical types of alkaline storage batteries in commercial use: one is the Hubbell, used in the miners' lamps of the Portable Electric Safety Light Co.; the other is the Edison storage battery.

214. The Hubbell battery differs from the Edison in principle only in using a negative plate of cadmium instead of the iron of the Edison battery. Hubbell apparently was the first to use nickel threads incorporated in the nickel-oxide active material, and this is one of the essential steps in the production of a practicable battery plate with this active material. The nickel oxide of the original Edison battery contained flake graphite to increase the conductivity; the graphite was seriously affected by electrolyte action, and these plates were short-lived. In the present type of Edison battery flake nickel replaces the graphite of the earlier type. In December, 1914, the first few of these batteries had completed 5 years of service.

215. Theory of the Edison battery. The active materials of the Edison battery consist of nickel peroxide for the positive plate and finely divided iron for the negative plate. The electrolyte is a 21 per cent. solution of potassium hydrate in water to which is added a small amount of lithium hydrate. To overcome the passivity of iron a certain amount of mercury is incorporated with the iron of the negative plate; a suitable compound is also incorporated with the nickel hydrate which is the salt from which the nickel peroxide is electrolytically formed. The nickel oxide is a relatively poor electrical conductor, and, for this reason, layers of flake nickel are added to the mass to increase its conductivity. The theory given by the Edison Storage Battery Company for the chief chemical reactions which take place in their cell after the first charge is:

$$3Fe + 8OH = Fe_3O_4 + 4H_2O \qquad \text{(Negative Plate)} \quad (11)$$

$$6NiO_2 + 8K + 4H_2O = 2Ni_3O_4 + 8KOH \qquad \text{(Positive Plate)} \quad (12)$$

$$8KOH + 6NiO_2 + 3Fe + 4H_2O = Fe_3O_4 + 2Ni_3O_4 + 4H_2O + 8KOH \quad \text{(Both Plates)}$$

$$6NiO_2 + 3Fe \underset{\leftarrow \text{charge}}{\overset{\text{discharge} \rightarrow}{}} Fe_3O_4 + 2Ni_3O_4 \quad (13)$$

A probably more correct theory is given by Dr. Fritz Foerster, Elektrochemie Wasseriger Losungen, p. 227, 1915, as follows:

$$Fe + Ni_2O_3 \cdot 1.2H_2O + 1.8H_2O \underset{\leftarrow \text{charge}}{\overset{\text{discharge} \rightarrow}{}} Fe(OH)_2 + 2Ni(OH)_2 \text{ (Both Plates)} \quad (14)$$

NiO_2 is first formed on charge but decomposes into the lower oxide, Ni_2O_3, according to the equation, $4NiO_2 = 2Ni_2O_3 + O_2$, giving off oxygen. This decomposition accounts for the continual gasing which takes place during charge and also for a considerable length of time after the charging is discontinued. If a cell is allowed to stand idle after charge, it will take from 1 to 2 days for the higher oxide to completely decompose to Ni_2O_3. This reduction is much hastened by discharge since any NiO_2 existing in the positive plate will be used up first. This rapid reduction on discharge of any NiO_2 which may be present probably accounts for the marked drop in voltage at the beginning of the discharge of a fresnly charged cell. Due to the presence of the higher oxide in the positive plate of a freshly charged cell, the voltage of such a cell is somewhat higher than the voltage of one that has been allowed to stand before discharge.

Foerster's equation shows that, although the potassium does not enter into the chemical reactions, the water of the electrolyte does and that there is an increase in the density of the electrolyte from the charged to discharged state. This change is opposite to the change which takes place in the density of the electrolyte of the ordinary lead accumulator. Although there is a decrease in the density of the electrolyte of an Edison cell during charge, this change cannot be used to determine the state of charge of the cell on account of the uncertainty of the amount of water lost by gasing during charge.

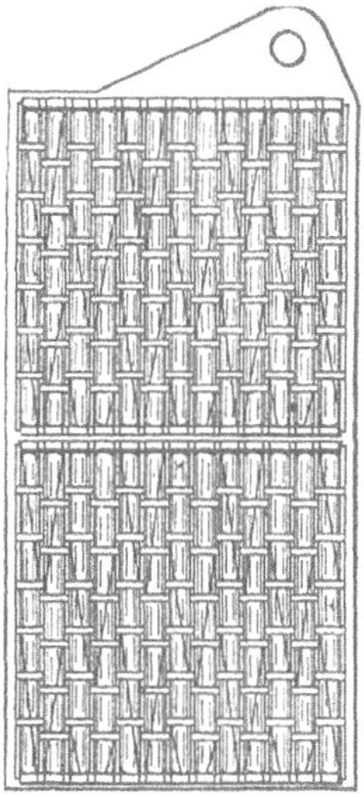

FIG. 51.—Edison positive plate.

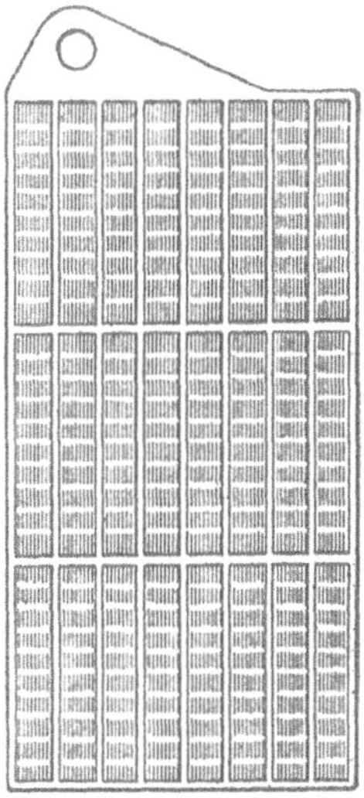

FIG. 52.—Edison negative plate.

216. Positive-plate construction. The positive plate consists of a nickel-plated steel frame into which are pressed perforated tubes filled with alternate layers of nickel hydrate and metallic nickel in very thin flakes. The tube is formed from a thin sheet of steel, nickel plated and perforated, and has a spirally lapped joint. The active material is tamped into the tubes, nickel hydrate and nickel flake being fed alternately and the tubes when filled have their ends pressed together in such a way to permit them to be mounted in the frames. The tubes have small steel bands slipped over them to prevent expansion away from the active material. Such a plate is shown in Fig. 51.

217. Data on Edison Cells

Type of cell	B-2	B-4	B-6	A-4	A-5	A-6	A-8	A-10	A-12
No. positive plates	2	4	6	4	5	6	8	10	12
No. negative plates	3	5	7	5	6	7	9	11	13
Weight (lb. per cell) without trays	4.7	7.5	10.8	13.9	16.7	19.4	27.2	33.3	41.3
Capacity (amp.-hr.)	37.5	75	112.5	150	187.5	225	300	375	450
Charge amp. for 7 hr.	7.5	15	22.5	30	37.5	45	60	75	90
Discharge amp. for 5 hr.	7.5	15	22.5	30	37.5	45	60	75	90
Avg. volts per cell discharged at above rate	1.2	1.2	1.2	1.2	1.2	1.2	1.2	1.2	1.2
Amp.-hr. efficiency	82	82	82	82	82	82	82	82	82
Watt-hr. efficiency	60	60	60	60	60	60	60	60	60
Watt-hr. per lb. of cell without trays	9.6	12.0	12.5	13.0	13.5	13.9	13.2	13.5	13.1

218. Negative-plate construction. The negative plate comprises a grid of nickeled steel with oblong openings into which are placed perforated steel boxes or pockets containing finely divided iron with mercury. See Fig. 52. These

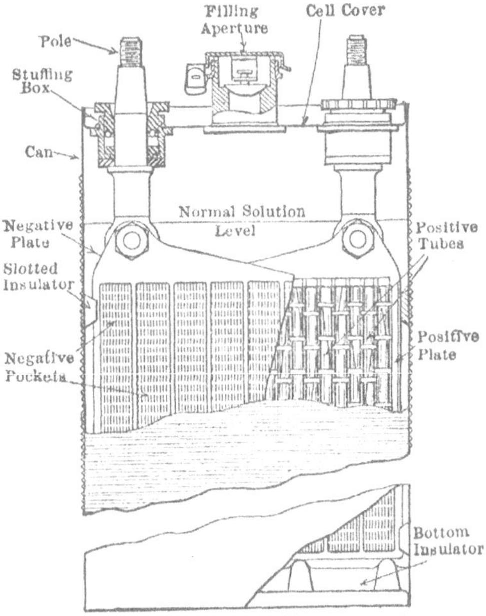

Fig. 53.—Edison cell construction.

boxes are assembled in the grid and subjected to pressure to weld the joints and to corrugate their surfaces. The iron is precipitated as a chemical compound, and the nickel hydrate and

Fig. 54.—Crate mounting.

this iron compound are converted electrolytically to nickel peroxide and metallic iron respectively, in the forming of plates.

219. The assembly of the cell is shown in Fig. 53. The plates are supported on hard-rubber spacing and insulating pieces. The lugs of the plates are punched and are mounted upon a steel pin with a terminal post. The ends of the pins are threaded and the plates, separated by washers, are held together by steel nuts. The elements are contained in a nickel-plated sheet-steel case, the walls of which are corrugated to add stiffness and also to assist in cooling the cells in action. The cover also is of nickel-plated sheet steel with three openings, two for the terminal bushings, which are provided with stuffing boxes, and the third for the gas vent and for the filling of the cell with water.

The cells are assembled in wooden crates as shown in Fig. 54, usually with the bottom left open to secure a circulation of air sufficient to keep the cells cool.

The normal rates of charge and discharge for Edison cells with A- and B-type plates are respectively 7 and 5 hours. The G-type plates are thinner than the A and B types and are for higher discharge rates. Cells of approximately the same dimensions with A- and G-type plates have the same ampere-hour capacity, but for a fixed voltage drop on discharge those with the G-type plates may be discharged at the higher rate. The normal rates of charge and discharge forlcells with G-type plates are respectively 4¾ and 3½ hours.

The numerals following the type designations in Table **214** indicate the number of positive plates. Each type-A positive plate has 30 tubes ¼ in. in diameter and is capable of giving 7.5 amp. for 5 hr. The type-B positive plates have half as many ¼-in. tubes and give 3.75 amp. for 5 hr. Cells with A-type plates are standard for electric commercial vehicles, trucks, locomotives, etc.

220. Charge and discharge curves. The characteristic normal charge and discharge curves for an Edison battery are given in Fig. 55. It will be noted that the average voltage on discharge is approximately 1.2 volts; the initial open-circuit voltage is approximately 1.5 volts, and the final voltage at the end of discharge is approximately 1 volt. The first part of charge shows a characteristic voltage rise, and there is a subsequent decrease in voltage with continued charge, followed by another rise. There is no extremely sharp rise in the final voltage of the Edison cell, and it is therefore somewhat difficult to determine the end of charge. If an Edison cell is

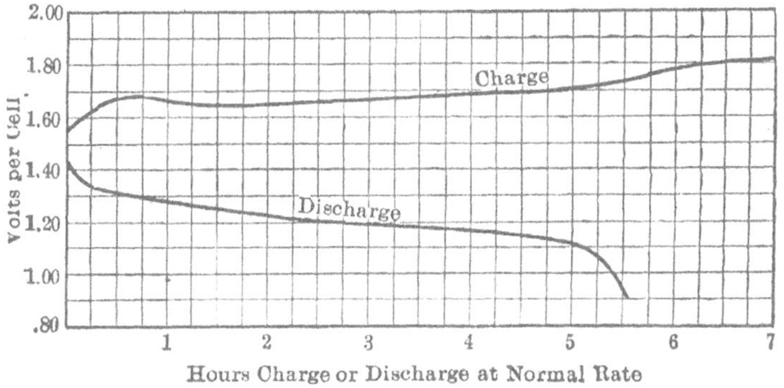

FIG. 55.—Normal charge and discharge curves.

known to have been considerably discharged, it is always well to continue the charging current for the full 7-hr. period, as the battery is not injured by overcharge unless the temperature passes a critical point.

221. Charging curves. Maximum and average voltages of charge as influenced by charging rate, are shown in Fig. 56. In these curves, the cell is assumed to have a constant temperature of 95 deg. Fahr. and the duration of charge is 7 hr. The voltage on charge decreases considerably with increasing temperatures.

222. Characteristic discharge curves. Characteristic discharge curves for Edison type "*A*" cells are given in Fig. 57. It will be noted that the ampere-hour capacity of an Edison cell does not suffer very greatly from increasing discharge rates, if the terminal voltage be carried low enough. For practical purposes, however, the last portion of the discharge at high rates would have no value because of its great falling off from the open-circuit voltage.

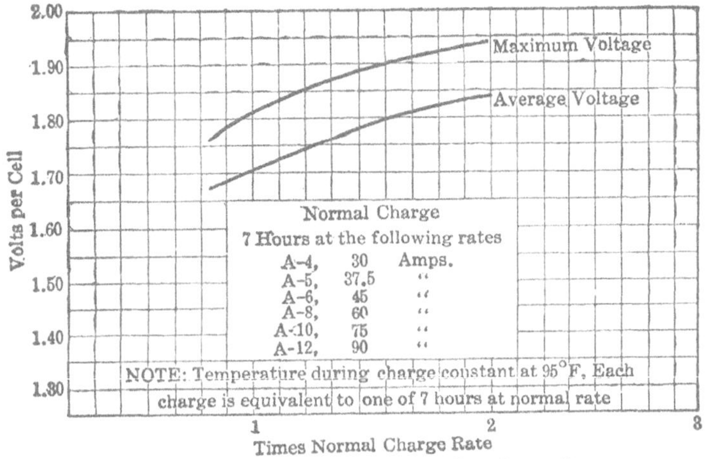

FIG. 56.—Temperature influence on charging voltage.

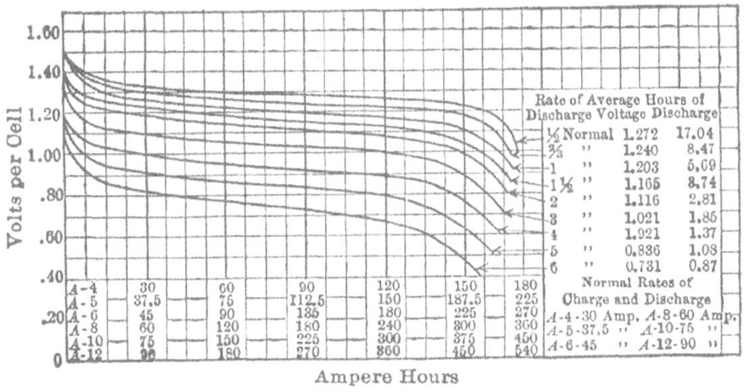

FIG. 57.—Typical discharge curves, type "*A*" cells.

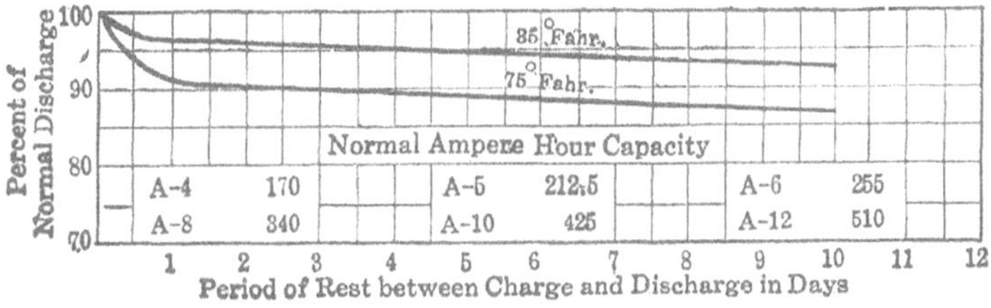

FIG. 58.—Loss of capacity.

223. Standing discharged. No injury is done to the Edison cell if it is completely discharged and allowed to stand in this condition. This is one of the principal distinguishing features of the Edison cell. The cell however, loses in capacity by standing as shown in Fig. 58, and is most easily brought back to its condition of full capacity by giving it a very considerable overcharge. The usual practice is to ship the Edison battery in a discharged condition, and it requires several cycles of charge and discharge to bring it to its full capacity.

224. Overcharge. The Edison battery is not injured by overcharging unless its temperature exceeds 105 to 110 deg. Fahr. High temperatures seriously affect the capacity and life of Edison negative plates, and the manufacturer's guarantee does not hold for overheating from this cause. A continuance of overcharge increases the capacity on the subsequent discharge, but this increase in capacity is obtained at the expense of efficiency. The effect of continued overcharge is shown in Fig. 59.

225. Effect of temperature. There is a marked falling off in capacity of the Edison battery with low temperature, especially at high discharge rates. There appears to be a fairly well defined temperature for any given discharge rate below which the capacity becomes quite suddenly reduced. (See W. E. Holland, Elec. Vehicle Assoc., Oct. 10, 1911.) For this reason the battery compartments of electric vehicles using Edison cells must be carefully insulated against cold, if low temperatures are to be met. If the battery has attained a low temperature from long standing, it will tend to heat up on discharge, because of the high internal resistance, and if this heat is retained by the insulation, a fairly complete discharge can be obtained.

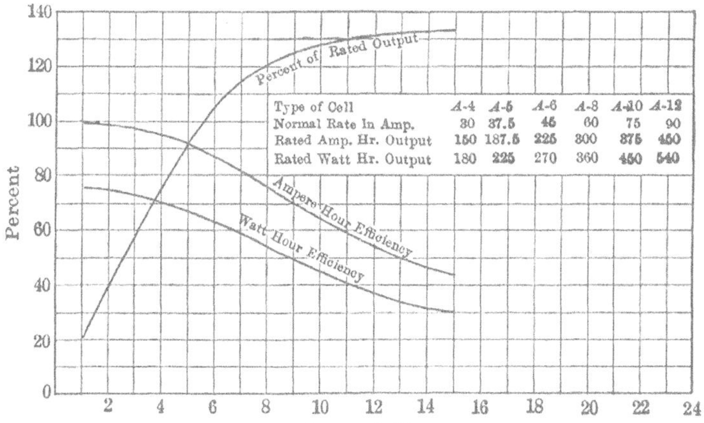

Fig. 59.—Influence of overcharge.

226. Deterioration of electrolyte. Water must be frequently added to the electrolyte of an Edison battery, to replace that lost by decomposition and evaporation. The amount of water required for this purpose is considerably greater than with a lead battery, because of the poor ampere-hour efficiency (due principally to the continuous gassing during charge) and the higher temperatures reached on charge. Distilled water must be used for this purpose, but even with its use, the electrolyte becomes gradually converted to potassium carbonate through the absorption of carbon dioxide from the air. For this reason also, it is preferable to use water which has been freshly distilled. As the electrolyte deteriorates from the above-mentioned cause, the capacity of the battery falls off, and, after about 250 cycles of charge and discharge, the electrolyte should be renewed. After each electrolyte renewal there is a temporary increase in capacity as shown in Fig. 60.

227. Testing. Tests on Edison batteries should be carried out as described in Par. **103** to **107**, with certain obvious changes. There is no appreciable variation of electrolyte density, and the end point of charge is also determined with greater difficulty than with the lead battery. A suitable neutral or test electrolyte for an Edison battery is a nickel-oxide tube, such as those used in the standard positive plates.

228. Application. The "*B*" type cells are used for ignition and lighting of gasoline motor cars; they are not commercially used for motor starting on account of the high internal resistance. The vehicle-type cells are used for electric vehicle propulsion, storage battery street cars, mining locomotives and industrial trucks; they are not used for load regulation on account of the heavy drop in voltage at high discharge rates and the high initial cost, these factors also limiting their use in other power applications.

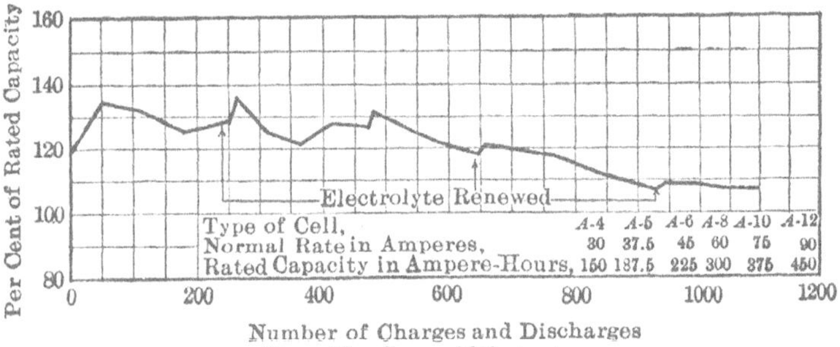

FIG. 60.—Log of life test.

229. Operation. The life of the Edison battery is guaranteed under certain restrictions as to its operation. The operating instructions of the Edison Storage Battery Co. should be carefully followed if the battery company is to be held to its guarantee. A temperature of 115 deg. Fahr. should not be exceeded under any circumstances, as high temperatures will seriously injure the negative plates.

230. Initial or first charge. Edison batteries are usually shipped in a discharged condition. Before placing them in service they should receive a continued charge at the normal or 5-hr. rate, for a period of 12 hr. or more.

231. Regular charge. If the battery has received a complete discharge, the charge should be started at the normal rate and continued for a period of 7 hr., or until each cell, under normal temperature conditions, has reached a voltage of at least 1.8 volts per cell. The ampere-hours of charge should exceed the ampere-hours of discharge by approximately 40 per cent., and the battery should receive in addition, an overcharge of several hours at the end of each month of service.

232. Replacing evaporation. The electrolyte must be kept well above the plates by adding water whenever necessary, to maintain the level. Always use distilled water for this purpose.

233. The outside of the cells and the trays must be kept clean and dry. Dampness under certain conditions will cause the containers to pit under electrolytic action.

234. Standing idle. If an Edison battery is to be placed out of commission it need not receive any special attention, other than to see that the electrolyte is brought to the proper level. The battery can stand either charged or discharged equally well. To obtain the full capacity, however, after a long period of standing it is necessary to overcharge the battery.

235. Life of Edison battery. A log of a life test as published by the Edison Storage Battery Co. is shown in Fig. 60. The statement is made that the conditions of the test were harder than would normally be met in service. The battery is more durable than the vehicle type of lead battery. It is doubtful if it approaches the durability of the heavy Planté type of lead cells.

BIBLIOGRAPHY

236. Selected reference literature on batteries.

COOPER, W. R.—"Primary Batteries." London, 1916.
Transactions American Electrochemical Society.
LEBLANC, M.—"Electrochemie." Leipzig, 1911.

PLANTE, G.—"The Storage of Electrical Energy." London, 1887.
SCHOOP, P.—"Die Secundare Elemente." Halle, 1895.
TREADWELL, A.—"The Storage Battery." New York, 1898.
ZACHARIAS, J.—"Die Akkumulatoren." Jena, 1901.
DOLEZALEK, F.—"Theory of the Lead Accumulator." New York, 1904.
JUMAU, L.—"L'Accumulateur Électrique." Paris, 1904.
LYNDON, L.—"Storage Battery Engineering." New York, 1911.
MORSE, H. W.—"Storage Batteries." Macmillan Co., 1912.
FOERSTER, F.—"Elektrochemie Wasseriger Losungen." Leipzig, 1915.
BUREAU OF STANDARDS.—"Electrical Characteristics of Dry Cells." Circular No. 79, April 25, 1919.
BUREAU OF STANDARDS.—"Cadmium Electrode for Storage Battery Testing." Technologic Paper No. 146, Dec. 12, 1919.
JUMAU, L.—"Étude Résumée des Accumulateurs Électriques." Paris, 1919.
Bulletins of the Battery Manufacturers.
Bibliography of References on Storage Batteries since 1899, compiled by G. S. MAYNARD, published by N. Y. Public Library in 1915.